MW01620771

ARCTIC WAR PLANES
ALASKA AVIATION
OF WORLD WAR II

Elmendorf AFB

Arctic War Planes
Alaska Aviation of WW II

A PICTORIAL HISTORY OF BUSH FLYING WITH THE MILITARY IN THE DEFENSE OF ALASKA AND NORTH AMERICA

By Stephen E. Mills

Bonanza Books • New York

Copyright © MCMLXXI by Superior Publishing Company
Library of Congress Catalog Card Number: 78-160190
All rights reserved.
Originally published as *Arctic War Birds.*
This edition is published by Bonanza Books
a division of Crown Publishers, Inc.
by arrangement with Superior Publishing Company
a b c d e f g h
Bonanza 1978

Reeve Collection

DEDICATION

TO ROBERT C. REEVE, WHO CONSIDERS HIMSELF fortunate to have arrived in Alaska in the embryo era of aviation. While others acknowledge him as a major factor in the history of the Northland, he credits the development of Alaska to aviation, and the growth of aviation to such pioneers as Carl Ben Eielson, Noel Wien, Harold Gillam, Art Woodley and Ray Petersen. "These men, and 100 more early flyers, actually opened up the Territory to settlement and commerce," Reeve said. "Without their guts, tenacity and ability, Alaska would never have become the hub of world travel."

Other major contributors to the industry were: U. S. Post Office and CAB Star Routes and later, certificates of scheduled carriers; the dozens of businessmen who invested their money in airplanes that were all too often left wrecked on some inadequate airfield; and to the State of Alaska who, with their modern attitude and willing cooperation, deserve credit for the successful progress of aviation when, in the earlier days, the Territorial Road Commissioner in Juneau viewed the airplane as a nuisance and did little or nothing to aid it.

AND to the thousands of Canadian and American servicemen and women who joined Alaskans in World War II to free the Aleutians of the invaders, and at the same time, put their stake in the future of Alaska.

Russell Photo

Admiral James S. Russell was born in Tacoma, Washington, on March 22, 1903. Graduating from that city's Stadium High School in 1918, he served as a merchant seaman during World War I. In June 1922, he entered the Naval Academy and graduated fifteenth in a class of 450 in 1926; received his Navy wings three years later at Pensacola, Florida. Postgraduated in Aeronautical Engineering, with the degree Master of Science in Aero Engineering from Cal Tech, Pasadena, in 1935. In World War II he commanded Naval Patrol Squadron 42, with service in Alaska and the Aleutian Islands; later served 18 months with the Bureau of Aeronautics in Washington, D.C. Returning to sea duty in June 1944 as Chief of Staff to Rear Admiral Ralph Davison, commander of Carrier Division Two in campaigns of Palau, the Philippines, Iwo Jima and Okinawa, he was sent to Japan immediately following the surrender for air technical intelligence. Later, following commands of the USS "Bairoko" and USS "Coral Sea," he was assigned to the Atomic Energy Commission staff. Admiral Russell shared with Mr. C. J. McCarthy the 1956 Collier Trophy for the development of the first carrier-based supersonic Crusader Navy fighter. From 1957 through 1961, he served first as Deputy Commander U. S. Atlantic Fleet as Vice Admiral and then Vice Chief of Naval Operations as four-star Admiral under Admiral Arliegh Burke, CNO. Before his retirement April 1, 1965, he had served as Commander in Chief to Allied Forces, Southern Europe, since 1962. He has served as consultant to The Boeing Company; director of Alaska Airlines and Airtronics, Inc., and on various Navy advisory boards. He was recalled to active duty in August 1967 to review safety in aircraft carrier operations and again in August 1968 as the chairman of a study committee in Southeast Asia for the Secretary of Defense.

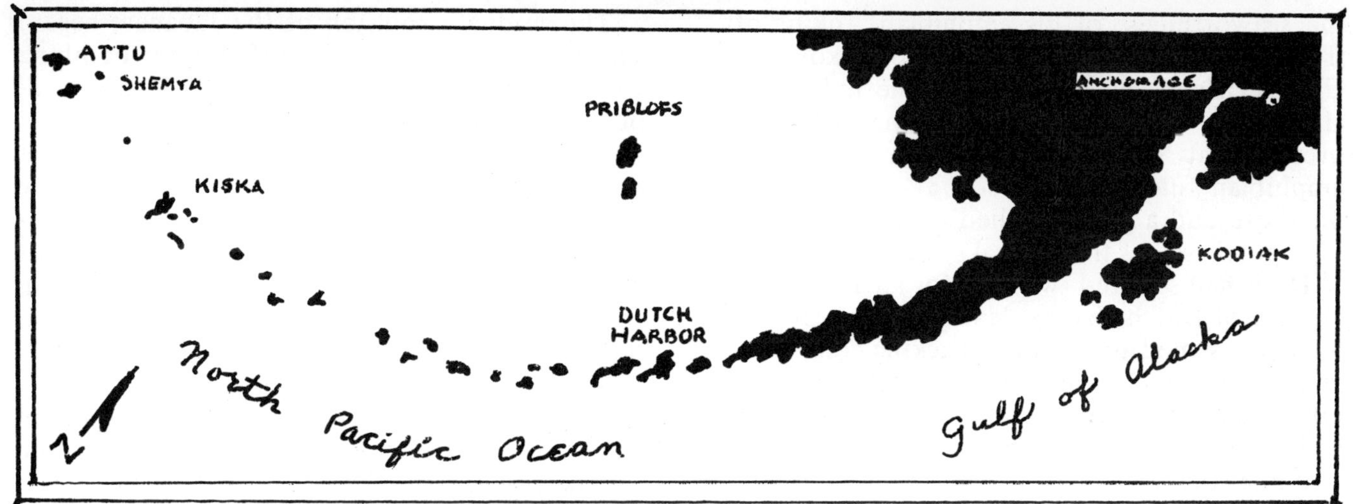

FOREWORD

by Admiral James S. Russell, USN, Ret.

MILITARY AVIATION was not unknown in Alaska prior to World War II, but compared to civil aviation, which brought the indispensable means of getting from here to there over difficult terrain, it was not very much before the public. Ladd Field was built with its underground "utilidor" connecting the various post activities as a place for the Army Air Corps to subject their airplanes to the rigors of winter in the Yukon basin. Fort Richardson and Elmendorf Air Base soon became the hub of furious Army activity. Fields suitable for landing bombers and fighters were begun seemingly everywhere – Annette, Yakutat, Whitehorse, Northway, Big Delta, Mile 26 (now Eielson), McGrath, Bethel, Galena, and Nome, to name a few.

The Navy, after aerial surveys of the tremendous extent of Alaska's shore lines, began seaplane bases at Sitka, Kodiak, and Dutch Harbor, installations which were only partially complete when the war began. Additionally, the Navy had based seaplanes on ships, called seaplane tenders, in various out-of-the-way places such as Kiska Harbor in the Aleutian Islands, Dolgoi Harbor in the Shumagins, and Cold Bay on the Alaskan Peninsula.

From Cold Bay, with an excellent harbor and a goodly stretch of flat land across the otherwise mountainous Alaskan Peninsula, to the rugged island of Attu is a distance of 850 nautical miles. Between these two points and arcing somewhat southward lie the Aleutian Islands. Volcanic in origin, extremely rugged with few good harbors and fewer level places for flying fields, bereft of any trees, these grim and forbidding islands are lashed by tremendous storms in the winter and bathed in fog, stratus clouds, and drizzle in the summer. The warmer water of the Pacific Ocean mixes with the colder water of the Bering Sea along this fringe of islands, and, in the mixing, generates foul weather. The islands were once inhabited by prosperous and resourceful natives, the Aleuts, to an estimated number of 10,000. However, fur traders, mostly from Russia, brought the white man's diseases – alcohol and barbarous treatment – which decimated the Aleut nation. The near extinction of the Aleuts was evident when I first flew down the Aleutian Islands in July of 1941. There were only two villages to the west of Nikolski on the island of Umnak. In a distance of 680 nautical miles, there were about forty natives who made their home on Atka, and sixty, under old Chief Mike Hodakof, on Attu.

In the summer of 1941, Major Everett Sanford Davis came through Kodiak in a Douglas amphibian. In his overnight stay, he soaked up all the information about the Aleutian Island which the Catalina pilots of Navy Patrol Squadron VP-42 could give him. Returning about a week later, he overflew Kodiak and went directly to Elmendorf for good reason – he doubted the seaworthiness of his amphibian. Taking off from a small lake near the village of Nikolski on Umnak Island, he had grooved the hull of his aircraft on a rock at the shore line. Anchored overnight at Sand Point in the Shumagins, the amphibian filled with water, but was drawn up on shore and a patch applied made from stovepipe purchased at the general store.

Davis had selected two sites for airfields in the vicinity of Dutch Harbor. Materials were shipped west, consigned to fictitious packing companies, and soon Marsden mat airfields began secretly taking shape at Otter Point on Unmak Island and Cold Bay on the Peninsula.

While the Japanese strike on Pearl Harbor made territorial inhabitants uneasy about Alaska's strategic and vulnerable position in the Pacific, the bombing of Dutch Harbor and subsequent occupation of Kiska and Attu shocked Alaskans into realizing that war was on their own doorstep.

Two things about the Aleutian Islands could be expected to excite the military curiosity of the Japanese. First, Kiska Island had a closed harbor, so declared by the U. S. Government a number of years before World War II. Kiska Harbor was "closed," for use only by the U. S., hence might be expected to be a target of the Japanese. The second thing about the Aleutians is that Adak, to the nearest degree, is on the same meridian as Midway Island in the central Pacific. The strike on Dutch Harbor, June 3, 1942 (west longitude date), the day before the strike on Midway, was an accepted military tactic representing as it did a feint on the north flank before the main attack in the center. A surmise that if Admiral Yamamoto had combined forces, adding JUNYO and RYUJO to the four carriers at Midway, the outcome there would have been "far different" is interesting. In our interrogation of Japanese officers in Japan at the end of the war, in which I was privileged to participate, they attributed the loss of their four carriers of the striking force at Midway to overconfidence and carelessness. The Japanese, however, in no way underrated the seriousness of their loss of the Battle of Midway.

After the Japanese occupation of Kiska and Attu, with the flexibility of forward basing on seaplane tenders, the Navy Catalina seaplanes and amphibians engaged in an early, furious bombing of Kiska, but this chore was soon turned over to the Army Air Force as Liberators became more plentiful in the theater.

The gradual expulsion of the Japanese began. Their position was made difficult by bombing which became more intense as U. S. forces occupied Adak and then Amchitka Islands. In May 1943, the bypassing of Kiska and the amphibious assault (the first island assault of the war by U. S. Infantry) on Attu set a pattern to be repeated many times over in the island-hopping of the Pacific. The bitter land action on Attu, from the initial landing of Willoughby's Scout Battalion to the suicide charge of Colonel Yamasaki with its last defenders, ended the bloody eighteen-day action. The Battle of the Pips on July 26 unwittingly cleared the way for the Japanese evacuation of Kiska on July 28, 1943.

The Aleutian campaign, the only World War II conflict on American soil, was the least publicized theater of operations. Limited in scope, it was, for its size, one of the costliest.

By the end of the war, military aviation had become an integral part of Alaska. It assisted greatly in the growth of Alaska to its present importance in air commerce.

Ellis Photo

Veteran early bush pilot Robert Edmund Ellis was recalled to serve in the Navy in World War II and in 1944-45, he commanded the Attu Navy sector.

Northwest Airlines

A. B. "Cot" Hayes, veteran Northwest Airlines' Alaska sales manager at Anchorage when NWA started Orient service in 1946, first came to Alaska in 1930 as division manager for Alaska Washington Airways and remained when that company was taken over by Alaska Southern Airways in 1932. Pan American Airways purchased ASA and Hayes stayed with PAA until 1938. He served a term as mayor of Juneau during 1944-45 and retired from NWA in 1962.

Western Airlines

Dave Kellogg earned his pilot's license in 1935 at Everett, Washington. In World War II he was an Air Corps' flight instructor before combat duty in the Pacific Theater, flying 55 sorties in a B-29. Capt. Kellogg returned to civilian flight instruction briefly before joining Woodley Airways as a bush pilot in the Kenai, Alaska area. In the early 1950s, he became a first pilot on Woodley's Pacific Northern Airlines' Anchorage-Seattle route and transisted to jet transports in 1963. PNA and Western Airlines merged in 1968 and Capt. Kellogg currently flies that firm's Anchorage-Seattle-Los Angeles flights. He lives in Seattle, Washington with his aviatrix wife, Terry, a member of the 99ers organization and past Powder Puff Derby participant.

ACKNOWLEDGEMENT

AT THE ONSET OF researching for *Arctic War Planes,* it was apparent that the limited documentation and availability of photographs of the World War II Aleutian Campaign warranted in-depth investigation. I wish to express my grateful appreciation to Bob Reeve and Admiral James S. Russell for the numerous contributions of valuable material and firsthand information related to me in many intriguing hours of interviews unselfishly granted from their busy schedules.

Additionally, rare photos and important information was generously contributed by Dr. John M. Weidman, Historian, Alaskan Air Command; Col. Harold S. Basham, USAF, Alaskan Command; Col. James E. Carter, CAP, Alaska Wing Commander; M/Sgt. Charles Lockhart, Historian, 21st Composite Wing; M/Sgt. Randall E. Hollingsworth, NCOIC, Public Information, Alaskan Air Command; whose cooperation, assistance and documentary material were of major importance.

Jack Stewart, shown at right during a 1935 interior Alaska hunting trip, came North in 1934. After a brief stopover at Anchorage, he moved to Fairbanks and joined the Pan American Airways communication department and manned PAA's first coastal radio station at Koyuk, Alaska. He transferred to Nome for five years until 1941 when he left Alaska to reside in Seattle, Washington.

Stewart Photo

Northwest Orient Airlines

Pretty stewardess Betty Gople is given a cockpit briefing of a Northwest Orient Airlines' Boeing 707 by veteran Captain Vincent Doyle. Now historian for Northwest, Doyle learned to fly at Minneapolis in 1939 and joined the airline in April 1943 after serving as flight instructor for the Air Corps. Assigned to the Korean Airlift in 1949, he flew Douglas DC-3 and DC-4 aircraft as captain. Doyle is currently checking out in the Boeing 747 jumbo jet and is an avid stamp collector.

A special appreciation goes to Vic and Marie Brown for their kind hospitality while researching in Anchorage; Al Horning, Jack Stewart, Don Sheldon, John Cross, Murrell Sasseen, Pat Wachel and Oscar Winchell, Dave Kellogg, Gil Cook, Donald E. Young, Ted Tax, Lillian Crosson Frizell, "Cot" Hayes and Polly Poyneer for photos and interviews.

Lastly, my heartfelt thanks to my lifelong friend and former co-author, James W. Phillips, who offered many hours of initial consultation and genuine encouragement; my wife, Pat Mills, for manuscript typing; and my sister-in-law, Sheri Howard, for over-all reading. With my sincere apologies to anyone overlooked, I have tried to cover this era as accurately as possible but I may have included some mistakes especially for those looking for them.

Boeing Co.

TABLE OF CONTENTS

U. S. Air Force

General John J. Pershing congratulating Capt. St. Clair Streett, who was in command of the July 1920 Billy Mitchell sponsored New York to Nome, Alaska Army survey flight. All four De Havilland DH-4 biplanes made the 4,320-mile round trip flight without major mishaps or injuries. Left to right: General Peyton C. March, Captain St. Clair Streett and General John J. Pershing.

PRE-WORLD WAR II

Alaska's Defenseless Days

CHAPTER 1

IN JULY 1913, at Fairbanks, Alaska, a New England airplane builder and pilot, James V. Martin, made the first powered flight in the Northern Territory. Martin, with the able assistance of his wife and a mechanic, introduced a new mode of transportation that was to play a major role in the development of Alaska.

Following token and experimental flights in and to Alaska from the "South 48," including an Army survey flight in 1920, the airplane was first put to commercial use shortly after World War I. Such aviation pioneers as Carl Ben Eielson, Noel Wien and Joe Crosson joined visionary businessmen in establishing preliminary air mail and passenger routes linking hub cities to key interior towns and way points.

The early 1930s witnessed a marked development in aircraft and aircraft engines, thereby advancing general aviation in the United States and giving birth to a large covey of new flyers. Many fledgling pilots and seasoned barnstormers alike migrated to the Northern wilderness of Alaska to stake their claim in aviation of the new country. Flying schools, maintenance shops, one-man airlines, along with home built pleasure aircraft, appeared on the scene almost overnight. The airports of Anchorage and Fairbanks became scenes of bitter competition vying for the scarce dollar. Alaskans were quickly adapting to the use of the airplane, but the demand for interior air travel fell far short of the availability of the air carriers.

Money was not overly abundant and many fares consisted of grubstaking a prospector; being paid when the trapper marketed his furs, and many interiorites would barter with food or services. Mercy flights were flown in the name of humanity, rather than cash payment. Aircraft were expensive to operate. Oil and gasoline were costly and often hard to come by in remote areas. Engine overhauls and repairs often meant

U. S. Signal Corps

Line-up of Army Alaskan flyers following completion of their historic 1920 New York to Nome survey flight. Left to right: Capt. Douglas, Lt. C. C. Nutt, Lt. R. H. Nelson, Lt. C. E. Crumrine, Lt. R. C. Fitzpatrick, Sgt. James Long and Sgt. Joseph E. English.

Lewis Collection

Upper: Three U. S. Army Air Service Douglas Cruiser biplanes, "Boston," "New Orleans" and "Chicago," ride at anchor near Seward, Alaska. A fourth plane, the "Seattle," crashed in Alaska shortly after the first world flight left Seattle April 6, 1924. The "Chicago" led the flight westward to its 26,345-mile completion at Seattle September 28, 1924; 363 hours, 7 minutes flying time later.

Noel Wien Collection

Noel Wien, standing at the tail of his World War I surplus Hisso Standard biplane in photo at left, made a successful landing and take-off from a 300-foot bar on Bear Paw Creek in the Kantishna mining area in the fall of 1924. In the summer of 1924 and 1925, Wien was the only commercial pilot in Alaska.

Group of Weeks Field, Fairbanks-based aircraft, pictured below in the summer of 1926, comprise the flying flotilla of Captain George Hubert Wilkins' first polar probe attempt from Alaska. Captain Ben Eielson was the expedition's chief pilot of the Fokker F-7 3M tri-motor in the center of the photo. Single-engine Fokker at left was destroyed in a take-off attempt. Veteran pilot Joe Crosson accompanied Wilkins and Eielson to Point Barrow in Swallow biplane.

Gordon Williams Collection

Bowers Collection

Above: Any small clearing often served as typical early day Alaskan airfield. However, this seemingly serviceable airport trapped the pilot into a laborious chore, ably assisted by a nearby farmer.

Photo at right, taken in the summer of 1929 on the shore of Cook Inlet near Kenai, Alaska, pictures Russell Merrill alighting from his Anchorage Transport Co. Travelair, which represented the first commercial air service of that city. In October, 1929, Merrill disappeared while on a flight to the Nyac Mine. Although fragments of the plane were found, Merrill's demise remains a mystery.

Stinson Detroit biplane and Hamilton all-metal monoplane of Wien Alaska Airways, in photo below, parked at Weeks Field, Fairbanks, in 1929. The Hamilton was later sold to Ben Eielson and is the same plane he and Earl Borland were killed in while on an Alaska-Siberia flight later that year.

Bob Gesell Collection

Photo by Noel Wien

Fairchild Hiller

Crosson Collection

Alaskan Airways' contract to remove valuable cargo of furs and supply provisions of the ice-bound schooner "Nanuk," pictured above, led to a famous saga of aviation history and the loss of Carl Ben Eielson, noted and decorated Arctic and Antarctic flyer.

Pioneer bush pilot and partner of Ben Eielson, Joe Crosson is pictured at left with his Alaskan Airways' Fairchild 71 he was flying when he and Harold Gillam found Eielson's wrecked Hamilton after seventy-seven days of searching.

Right: Carl Ben Eielson, who made the first official air mail flight in Alaska on February 21, 1924.

Crosson Collection

Enroute from Fairbanks to Nome to join in the Eielson search, pioneer Canadian bush pilot T. M. "Pat" Reid, caught in a snowstorm, was forced to make an emergency landing resulting in damage to the Fairchild's starboard wing tip. Below, mechanics Hughes and Hutchinson make temporary repairs in sub-zero temperatures. The downed flyers were marooned a total of six days.

Fairchild Hiller

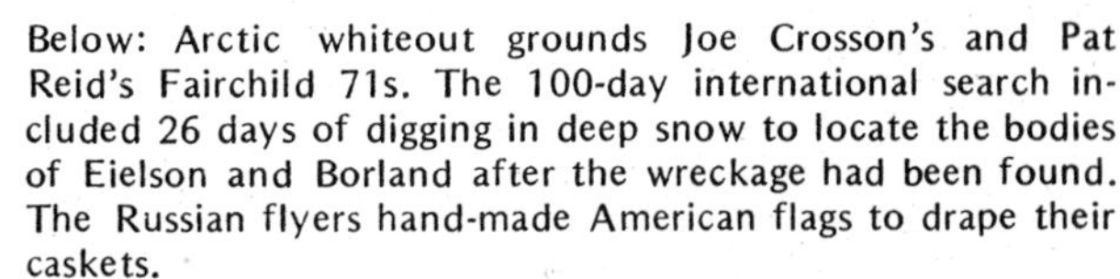

Below: Arctic whiteout grounds Joe Crosson's and Pat Reid's Fairchild 71s. The 100-day international search included 26 days of digging in deep snow to locate the bodies of Eielson and Borland after the wreckage had been found. The Russian flyers hand-made American flags to drape their caskets.

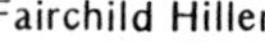

Fairchild Hiller

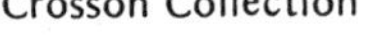

Crosson Collection

Above: Commander Slipnev flying one of two Russian Junkers used on the Eielson search lands at Teller, Alaska.

Left: Harold Gillam, who with Joe Crosson, found Eielson's crashed Hamilton monoplane.

Parka-clad Marian Swenson, daughter of the "Nanuk's" owner, and Pat Reid pause between search flights for photo at right.

Crosson Collection

Fairchild Hiller

Pilot Harold Gillam's Stearman search plane, below, is greeted by anxious "Nanuk" crew members and search-base personnel. Gillam, a fledgling pilot, demonstrated on the Eielson search his uncanny ability to "punch holes in the weather." He and Joe Crosson found the Hamilton wreckage on January 25, 1930.

Below: Wreckage of Eielson's plane on Siberian shore is almost covered with snow after being down over two months. Joe Crosson and Harold Gillam spotted the ill-fated plane when the sun reflected off the metal wing shown protruding in the center of photo. Eielson and his mechanic, Earl Borland, were thrown clear in the crash necessitating hunting for their bodies in deep snow.

Fairchild Hiller

Crosson Collection

Boeing 204 flying boat of Gorst Air Transport was operated out of Cordova in 1929 by general manager Clayton L. Scott. A sister ship was lost when engine failure forced Scott to make a rough surf landing in the Gulf of Alaska. Scott and co-pilot Gordon Graham, flying a Keystone Loening, made the first commercial flight across the Gulf of Alaska from Juneau to Cordova in 1929.

Gordon Williams

shipment to repair facilities "outside." Damaged airplanes could mean being out of business or suspended operations until parts could be shipped in and repairs completed.

Another deterrent in the Territory, foul flying weather, caused more than one enthusiastic potential airline owner to go bankrupt. Adverse weather could keep planes grounded for days and sometimes weeks. If fair weather prevailed after the difficult freeze-up and thaw seasons, anxious pilots flew almost every waking hour in an effort to recoup their losses. Tired engines and overloading occurred more than was generally known.

Lucrative and steady accounts, like fur buyers, diverted for the most part to the safe, cautious, dependable pilot-operator, of which there were many. Some unwanted, hazardous accounts fell to more skilled pilots who accepted the challenges in order to survive in business. The rugged operating conditions in the Territory took its toll in life, aircraft and finances, and soon only the experienced, shrewd and lucky became established in aviation commerce.

During this period of aviation development in Alaska, their neighbors, the Canadians, achieved a parallel state of aviation progress and a close comradery developed between airmen of both countries which manifested itself on many searches, for flyers of both countries participated in equal intensity.

Pioneer bush pilots of Alaska developed a realistic, panoramic concept of the magnificence of their country. Adversities notwithstanding, being members of a fraternity not earthbound, they were the first from the beginning of time to be visually aware of the Territory's vastness, potential and necessity for air travel development. The magnitude of isolated mineral-rich deposits, thousands of miles of virgin timber, unmolested wildlife and inherited pioneer spirit, nutured most "sourdoughs" into a possessive independence and hospitably sharing their domain with all newcomers or "cheechakos," except the exploiter and the undesirable.

Throughout the 1930s, the airplane burst from the slow, water-cooled engined wood and fabric biplanes to high-powered, closed cabin all-metal aircraft capable of speeds four times their World War I ancestors. The extended ranges invited highly competitive international endurance and air races and around-the-world record breaking flights. While shrinking the globe, airplanes made accessible the remotest areas, including the North and South Poles. Captain George Hubert Wilkins, with pilot Carl Ben Eielson, made a suc-

Fairchild Hiller

Alaska Washington Airways' Fairchild .71 taxies on Lake Union, Seattle, preparatory to a flight to Ketchikan, Alaska. In 1932, the bankrupt AWA was purchased from J. L. Carman by early aviation promoter Nick Bez, who renamed the firm Alaska Southern Airways. Bez sold the airline to Pan Am Airways in 1934.

Below: Pacific Alaska Airways, a subsidiary of Pan American Airways, flew Fairchild 71s as standard equipment from Fairbanks and Juneau during the middle 1930s. Sled dog, on the river bank observing loading operations, epitomizes the airplane's assumption of one of the former modes of interior transportation.

Pan American Airways

Alaska Airlines

The arrival of the airplane was a big event in most Eskimo villages. Many bush planes operated on pontoon gear during the summer with lakes, rivers and salt water offering access to much of the Alaska interior.

Pioneer pilot Noel Wien lands two prospectors at their campsite on the east fork of the Chandler River, 240-miles north of Fairbanks. Prospectors and trappers became steady customers of the bush flyer, and most were good pay.

Photo by Noel Wien

Bob Gesell Photo

Vic Ross of Fairbanks takes to the air in his Northern Air Transport Stinson SM2A at Kotzebue, in photo at left. Kotzebue, by early standards, had a better-than-average airfield. The present airfield at Kotzebue, Wien Field, is name for Ralph Wien, who was killed in a plane crash there in October, 1930.

At right, natives anxiously await a bush plane to take their stricken son to the hospital at Fairbanks. Mercy flights were common and the availability of the airplane saved hundreds of lives. More than one bush plane lost a race with the stork and the pilot served as midwife.

Alaska Airlines

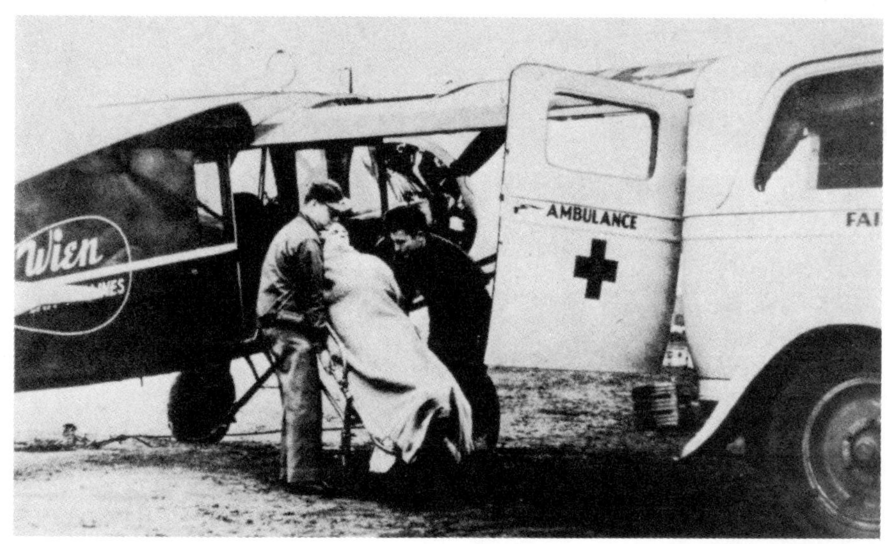

Wien Consolidated

The bush pilot's duties were many and varied; pilot, mechanic, bookkeeper, salesman, swamper and ambulance assistant, as shown in photo at left in Fairbanks.

Gordon Williams

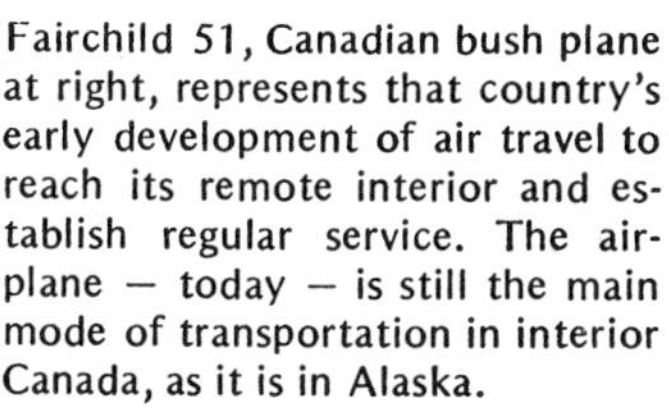

Fairchild 51, Canadian bush plane at right, represents that country's early development of air travel to reach its remote interior and establish regular service. The airplane — today — is still the main mode of transportation in interior Canada, as it is in Alaska.

The Lockheed Vega, "Winnie Mae," with pilot Wiley Post and navigator Harold Gatty, arrives at Fairbanks in July 1931 on the last leg of their around-the-world flight. Mechanics of Joe Crosson's Alaskan Airways serviced the fast monoplane, including changing a damaged propeller bent in a taxi accident on the beach near Solomon, Alaska. Post temporarily straightened the prop with a rock and a wrench, enabling the pair to continue their famous endurance flight.

Young Photo

cessful fixed-wing flight over the North Pole from Point Barrow, Alaska to Spitsbergen, Norway, on April 16, 1928.

Wiley Post made two successful around-the-world flights; first in 1931 with Harold Gatty, and again solo in 1933. Both flights necessitated landings in Alaska. These events, plus many others, focused world attention on the advance of aircraft and Alaska.

The rapid progress of the airplane and air commerce prompted the U. S. Government to accellerate departmental expansion to keep pace with a newly created commerce. Originally under the Department of the Interior, regulation and control of the civil air industry and all its

L. "Mac" McGee came to Alaska in 1929 in search of furs and wound up in the airline business. Interiorite demands for use of his first plane, while contacting trappers for furs, mushroomed McGee Airways into one of the largest operators out of Anchorage in the early 1930s. Remote village stopover of McGee's Stinson SM8A, pictured below, typifies Alaska's new mode of transportation.

Alaska Airlines

Mills Collection

Alaska's first flying school, Star Air Service, was established at Merrill Field, Anchorage, in March 1932. Charles Ruttan, Jack Waterworth and Steve Mills, partners in the venture, taught many Alaskans to fly in Fleet biplane and Curtiss Robin in upper photo. The Star soon branched out into commercial aviation and, with McGee Airways, were the founders of Alaska Airlines.

Below: Mary Barrows became the first licensed woman pilot trained in Alaska shortly after this July 1932 photo taken with her instructor, Steve Mills, and Fleet trainer.

Barrows Photo

Hubbard Collection

Percy Hubbard, owner of Service Airways in Fairbanks, and Ralph Wien both learned to fly in Alaska. Wien started his career in the North country as a mechanic for his brother, Noel, and after learning to fly in 1928 became one of the region's best-known flyers.

Reeve Photo

Above 1933 photo is of Bob Reeve and his Fairchild 51 at Valdez, Alaska, where he first established Reeve Airways. Reeve migrated to the Territory in 1932 after more than three years of Andes Mountain flying for Pan American Grace Airways of South America. He was one of the few experienced multi-engine pilots in Alaska at that time.

Art Woodley is shown below as he pushes his pontoon-equipped Travelair out from the shore of Lake Spenard. Woodley came to Anchorage in 1932 and established Woodley Airways and built that line up to become Pacific Northern Airlines after World War II. He and General Curtis LeMay were flying cadet classmates in 1929.

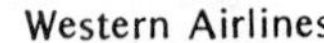
Western Airlines

Pan American Airways

Part of the fleet of Pacific Alaska Airways Fairbanks-based Fairchilds and their support crews are shown here in this 1934 photo, above. Standing, left to right: Larry Davis, Warren Tillman, Dick Ortman, Frank Panning, Jack Warren, Gordon Springbit and Andy Hufford. Kneeling, left to right: Jinx Ames, Al Monsen, Jack Egan, Kelly and Walt O'Toole. Note oversized "donut" tires for rough interior landing areas.

Alaska Airlines

Right: Self-photo of bush pilot Estol Call of McGee Airways in 1934. Call made this emergency camp when forced down by weather in the interior. Photography was Call's hobby; as well as Noel Wien, who took the largest number of excellent early day aviation pictures.

U. S. Air Force

Ten Martin B-10 bombers fly in formation over Washington, D.C. winging their way to Fairbanks, Alaska. Led by Col. Hap Arnold, the Army survey flyers, in addition to mapping assignments and photographic missions, spent time with bush pilots learning the problems of Northern flying and their techniques for cold-weather operations.

Below: Line-up of B-10 bombers at Fairbanks, Alaska, during July through August 1934 Alaska flight. The occasion gave many Alaskans their first view of military air power.

U. S. Air Force

James Hartley Photo

Above: Two Army Douglas camera biplanes, "Tennessee," left, and "Michigan," right, that accompanied the bombers on their Alaskan tour are temporarily based at Merrill Field.

Lt. Col. Hap Arnold and his executive officer, Major Hugh Knerr of the Alaskan flight, go over map with their projected route prior to take-off from Fairbanks.

Below: Alaskan flight officers gather for briefing before leaving Alaska for Washington, D.C.

National Archives

Crosson Collection

Gordon Williams Photo

Wiley Post's Lockheed Vega, "Winnie Mae," is shown in upper 1933 photo at Floyd Bennett Field being serviced for his proposed around-the-world solo flight. Post intended to beat the previous global speed record established by him and navigator Harold Gatty in July 1931. The Vega was powered by a Wasp Super.

Post landed his high-speed Vega on the short airfield at Flat, Alaska, resulting in the collapse of the right landing gear in a drainage ditch. The minor damage was repaired while Post slept. Bush pilot Joe Crosson led Post to Fairbanks where the "Winnie Mae" was refueled. The famous endurance flyer continued on to New York, breaking his previous record by 21 hours.

Waterworth Photo

Mills Collection

Bad flying weather, rough terrain and engine failure were just a few of the factors involved in crashes and accidents. Bellanca, in above photo, caught fire while flying in subzero temperatures through a mountain pass. Only one wing was intact when the flames were extinguished. The six-place monoplane was patched up at the remote interior location and flown to Anchorage where it was completely rebuilt.

Pontoon-equipped Curtiss Robin, at right, folded up when pilot hit shore of small lake he attempted to land in. Aircraft was put back in service in less than a week.

Mills Collection

Kellogg Photo

Engine failure forced the pilot of this Travelair monoplane, pictured at left, to make an emergency landing on a bog which tripped the aircraft, throwing it on its back. After minor repairs, the plane was flown out on skis after the bog became frozen.

Two fatalities resulted when the pilot of this new Ryan monoplane found himself in a whiteout condition; where the snow covered landscape blends with a murky sky. The ship slammed into the frozen lake at full speed.

Mills Collection

Gordon Williams Photo

Above: Larger aircraft, like this all-metal Ford 5 ATC tri-motor, began operating in Alaska about 1935. The Ford's rugged construction and additional engines made possible heavy payloads, such as mining equipment, supplies, and gasoline drums for remote fuel caches. With one exception, the Fords flown in Alaska were restricted to ski and wheel operations.

Small, fast Monocoupe of Johnny Littley's Alaskan Air Service, in 1934 photo at right taken in Anchorage, represents an example of the numerous modern planes that were appearing on the Alaskan scene. William Egan, twice-elected Governor of Alaska, learned to fly in a Monocoupe at Valdez in 1932.

James Hartley Photo

Gordon Williams Photo

Pacific Alaska Airways Ford Tri-motor, sans two outboard engines, flew throughout all of Alaska and was one of the few Fords fitted for pontoon operations. Note the engine is canted slightly downward for more efficient single-engine operation.

Hans Mirow's Sikorsky S-39-B amphibian, shown at left, was one of the varied type of aircraft flown in the Northland. Flying boat configurations were popular in southeastern Alaska where airports were limited.

Fahlin Plymocoupe, dubbed "Seaska," crash-landed short of the Juneau airport when enterprising pilot-owner Russ Owen thought he had lost the oil pressure in the aircraft's 1935 Plymouth engine. Owen promoted backers for his proposed Anchorage to Seattle non-stop flight in 1936 to prove the endurance of automobile engines for light plane use. Investigation of the accident revealed the engine was not at fault, but had a defective guage.

Lower: Pacific Alaska Airways inaugerated Lockheed Electra service in Alaska in 1935 as preparation for trunk-line schedules from Fairbanks to Juneau.

Gordon Williams Collection (Photo by Al Henderson)

Stewart Collection

Left: Irving McGuire, Ray Petersen and Jack Stewart came to Anchorage in 1934. McGuire and Stewart soon moved to Fairbanks. Petersen flew briefly for Star Airways; then established Ray Petersen Flying Service at Bethel, which later became a part of Northern Consolidated Airlines with Petersen as president.

Al Monsen came North in 1925 and worked for the Alaska Railroad. He learned to fly in Oakland, California, in 1928 and first flew for Pacific International Airways out of Anchorage with Harry Blunt, Alonzo Cope, Frank Dorbandt and Matt Nemienen. He later joined Pacific Alaska Airways and is shown below at Koyuk, a PAA stop between Fairbanks and Nome. Monsen earned a wide reputation as a safe, dependable, above-average pilot.

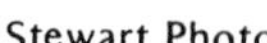
Stewart Photo

Right: PAA radio operator Jack Stewart, at Koyuk in 1935, holds large spider crab with local Eskimo woman who acquired the nickname "Short and Dirty" when on a flight to Nome the ticket agent, unable to decipher her Eskimo name, aptly described her on the passenger manifest.

Lower Right: PAA, one of the first bush operators to develop ground-to-air communications, used a sod roof log cabin at Koyuk as their first station out of Fairbanks. The single-windowed shanty with cracks in the floor, dirt from the ceiling and small, pot-bellied stove, made Jack Stewart's first season less than comfortable. Battery-operated transmitter was recharged at night. Bob Gleason, Vic Page, Oxel Johnson, Charles Huntley, Al Lane and Bruce Hensley were some of the early PAA interior radio operators.

Stewart

Below: Eskimo family and neighbors of the PAA radio station at Koyuk, located on the northern shore of Norton Sound at the mouth of the Koyuk River. The village population consisted of a white school teacher and her husband, 100 Eskimos and 500 dogs.

Stewart Photo

Stewart Photo

Gordon Williams Photo

Left: Wien Airways' Bellanca in which pilot Noel Wien with Vic Ross flew first commercial Fairbanks to Seattle flight in August, 1935, with passenger Alfred Lomen and first Post-Rogers crash photos taken by Emil Jacobs.

Gordon Williams Photo

Above: PAA Lockheed Electra with crew members Joe Crosson, pilot, Bill Knox, co-pilot and Bob Gleason, radio operator, lands at Boeing Field (Seattle) from Fairbanks with bodies of Post and Rogers on August 19, 1935.

Gordon Williams Photo

Left: PAA Douglas DC-2 funeral plane, with victims of the crash that shocked the world, departs Boeing Field on flight to California.

Former PAA Fairchild 71, at right, was purchased by veteran pilot Alex Holden and boat operator Jim Davis to form Marine Airways of Juneau in 1936. The popular Fairchild 51 and 71 aircraft were designed and built by Sherman Fairchild for aerial photography and developed into widespread demand for rugged Alaskan and Canadian use.

Gordon Williams Photo

Gordon Williams Photo

Bellanca Skyrocket of Alaska Air Transport based at Juneau was owned by Shell Simmons, who founded the company in 1935. Simmons was the sole operator out of the capital city when PAA pulled out of southeastern Alaska later that year. The forming of the competitive Marine Airways the following year resulted in a merger in 1939 to become Alaska Coastal Airlines.

Numerous, and some unusual, airplanes began to appear on the Alaskan scene, as in the case of Tony Schwamm's Italian-built Savoia Marchetti S-55. Schwamm, a former Hollywood stunt pilot, purchased two of the flying boats, one for parts, to establish a Petersburg-Seattle air service. The plane was destroyed as a result of a mooring accident and subsequent sinking.

Gordon Williams Photo

Gordon Williams Photo

Wright-powered CH300 Bellanca, of Herb Munter's Aircraft Charter Service, taxies to a mooring ramp following a flight from Ketchikan to Seattle in April 1936. Munter, who built and flew his own plane at age fifteen, joined many Northwest aviation pioneers attempting to establish airlines to Alaska.

Munter and Herb Munter, Jr., are pictured at left in 1938 at Hidden Inlet, Alaska. Munter, Jr., as a U. S. Navy pilot, was killed during World War II as a result of a mid-air collision while flying formation in foul weather near the coast of Washington.

Monro Stickney Photo

Gordon Williams Photo

Stinson SR-10 Reliant, right, of Pacific Alaska Airways, was used by staff personnel during PAA's transition from bush operations to trunkline routes to Alaska.

Webb Collection

Ray Webb, of Seattle, flew repairmen and parts to southeastern Alaska canneries for American and Continental Can Companies in 1938. Occasionally, he flew at wave-top heights and around islands when the weather was adverse. Webb flew his amphibian biplane in stunt scenes for the 1938 Paramount cinema, "Men with Wings," doubling for the star, Ray Milland. The rare Kinner-powered Budd "Pioneer" was an American stainless steel version of the Italian-built all-wood Savoia Marchetti SM-56. Webb eventually returned to flying out of Westport, Washington, until his death of natural causes in 1960.

Howard Hughes, with his fast, modern Lockheed 14-W Special shown below, circumnavigated the world in three days, eight minutes and ten seconds, establishing a new record. The July 1938, 14,825-mile flight placed one refueling stop at Fairbanks, Alaska, and Hughes, like Wiley Post, realized the importance of Alaska in world air travel. At Fairbanks, veteran Alaska pilot Joe Crosson listened as Hughes related in airman's jargon his experiences of flying over vast, unknown and uncharted country. Originally, Hughes had planned to use a Douglas DC-2 on the famous flight, but switched at the last minute.

Gordon Williams Collection

Veteran bush pilot Oscar Winchell of McGrath, Alaska, inspects a Cessna Airmaster, one of the newer, modern aircraft appearing in the Northland about 1940. His talented daughter, Pat Wachel, in photo at left with Winchell, is the authoress of her father's biography, "Alaska's Flying Cowboy," published in 1967.

Winchell Collection

The large number of aircraft arriving in Alaska just prior to World War II is evident in 1941 photo, above, of the airport at McGrath. McGrath, an important trading and supply junction in pre-airplane days, continued the role with air travelers.

Stewart Collection

Below: Nome-based Lockheed Vega of the Mirow Air Service is being dismantled for shipment and repairs following a landing accident on rough ice. Owner Hans Mirow was killed in 1940 while searching for one of his lost pilots, Frederick Chambers.

components fell to the Civil Aeronautics Authority (CAA) on August 22, 1938. Broadening the scope of safety regulation, the CAA also was authorized to subsidize air carriers in the form of mail pay. For the most part, a marked degree of bitterness and resentment by Alaska's bush pilots greeted early CAA inspectors and authorities. Their struggle for existence was difficult and the feeling was that Stateside "by the book" requirements were not that applicable to problems and conditions faced by the Northern counterpart. The CAA and the Territory's air carriers, after many high-level skirmishes, eventually found equal ground and meanwhile, the other half of the government agency was making its awareness in the form of charts, navigation aids, airports planned and built, and route scheduling and allocation. By the end of the 1930s, aviation in Alaska had become an established vital segment of air commerce.

By 1940, with newer planes, pioneer bush pilots, grubstakers and businessmen advanced Alaska air travel by establishing air mail and scheduled commercial routes between key populated areas. Remote towns and villages enjoyed home-based, one-man or small bush operations. Many latecomer pilots, not wishing to join the fast developing larger companies, answered to the demand of small, remote airplane business. They learned firsthand early aviation experience, hardships and disasters that, when heeded, allowed them to serve the isolated trapper, lonely prospector and dependent village.

With the airplane in service to the most distant village, news of the dark war clouds in Europe became a matter of concern. That total global conflict was a possibility soon became apparent. It was with this alarming thought that the U. S. Congress focused their attention on Alaska and its close proximity to Asia and vulnerability by a foreign power.

Commercial aviation in Alaska, although established, had not fully come of age in the dawn of the 1940s when, through meager congressional appropriations, the Army moved into Fairbanks and Anchorage. These cities suddenly viewed the servicemen everywhere amid the sounds of military bases under construction.

A nostalgic, pioneering era of the Territory and colorful aviation development was about to close.

Below: 1941 photo of Pacific Alaska Airways Lockheed Electra at Ruby, Alaska, symbolizes the end of the pioneer bush pilot era and the establishment of modern scheduled service to most key points throughout the Territory. For the most part, 100 airmen contributed to the introduction of aviation to the Northland up to 1941. Some are still active in one of Alaska's most important industries today — aviation.
Stewart Collection

Elmendorf Archives

In pre-Pearl Harbor photo above at Ladd Field, Fairbanks, Lt. Gen. John L. DeWitt, Commanding General of the Fourth Army and the Western Defense Command at San Francisco, center, confers with Brig. Gen. Simon Buckner, Jr., Alaska Defense Commander, right, and Colonel Dale V. Gaffney, Commanding Officer of Ladd Field and cold-weather station coordinator, on the progress of defense building in Alaska. Until mid-1940, Alaska had never had a defense commander. DeWitt picked Buckner for the job.

SECRECT AIR BASE BUILD-UP

General Simon Buckner's Task

CHAPTER 2

EARLY IN 1940, despite the pleadings of Alaska's Delegate to the U. S. Congress Anthony J. Dimond to defend Alaska, Federal appropriations for that year approved $4,000,000 for a military installation near Fairbanks, but as a cold-weather station rather than for defense. Thus, construction of Ladd Field (now Fort Wainwright) began.

The appropriations bill was voted and passed on April 4, 1940, despite urgings of General George Marshall and Major General Hap Arnold to include funding for the establishment of an operating air base at Anchorage.

Shortly thereafter, Adolph Hitler intensified Germany's conquests in Europe and Operation Orange, one of the rainbow plans for U. S. Hemisphere Defense, was activated under Lt. General John L. DeWitt, commanding general of the Fourth Army and the Western Defense Command at San Francisco. The plan called for a triangular defense; Panama, Hawaii and Alaska, for the possibility of war against the military regime of Japan.

The U. S. Senate quickly approved additional appropriations for air bases at Anchorage, Kodiak, Fairbanks, Yakutat, Annette Island and a naval base at Dutch Harbor.

In April 1940, Alaska defense forces of 774 enlisted men and 30 officers of the 4th Infantry arrived at the tent site of what was to become Fort Richardson at Anchorage. On July 22, 1940, Alaska's first defense commander, Colonel Simon Bolivar Buckner, Jr., arrived at Anchorage to assume Alaska's military protection against possible conflict. To accomplish his mission, Buckner had to build his northern defense force from scratch, creating garrisons, airfields, communications and supply lines. This was comparable to carving a civilization out of

Below: The first Air Corps' flight personnel assigned to Alaska's defense, Maj. Everett S. Davis, Sgt. Grady and Cpl. Smith, landed on the grass runway of Anchorage's Merrill Field August 12, 1940, in an obsolete Martin B-10 bomber. Davis made his temporary headquarters at the civilian airport pending completion of facilities and runway at Elmendorf Field. In March 1941, Davis was designated Chief of Aviation, Alaska Defense Command.

James Hartley Photo

the wilderness, compounded by the fact that practically all supplies, including food, had to be transported from Seattle by ship.

With war in the Pacific a possibility, and in Buckner's mind a probability, he was completely dismayed that Alaskans, the most air-minded people on earth, had no Air Force. For the past decade and a half, bush pilots had become the life link to inhabitants of over 150,000 square miles of America's northern possession and such aviation greats as Wiley Post, Howard Hughes, Billy Mitchell, Ben Eielson, Charles Lindbergh and General Hap Arnold had flown there and recognized the strategic importance of Alaska. The U. S. Congress not only had not recognized this, but when Buckner arrived, no money had yet been released and virtually no manpower or materials were available. With Dimond and Arnold working on Washington to unsnarl the tangled funds, big Kentucky-born Buckner employed every means conceivable to get the job done in the shortest possible time.

Two early, strong advocates of Buckner's task were Colonel Dale Gaffney at the cold-weather station in Fairbanks and newly-arrived Major Everett Sanford Davis, who set up the first Eleventh Air Force headquarters at Anchorage's Merrill Field in a wanigan and proceeded to conquer the multitude of problems involved in setting up air operations.

While Davis worked closely with Gaffney and bush pilot veterans like Joe Crosson and others on cold-weather operations and procedures, Buckner hired other bush pilots like Bob Reeve to haul supplies and materials once they arrived from "outside" and stockpiled at the docks. Roads were non-existent and the Seward to Fairbanks overworked Alaska Railroad could not handle it.

By late 1940, the shrewd now Brigadier General Buckner and his staff had created a thin spider web of military protection throughout the center of Alaska. His attention now turned to the Aleutians.

The summer of 1941, now Lt. Colonel Davis made a survey out to the end of the Alaska Peninsula and into the chain surveying for areas suitable for air bases. With sites picked at Cold Bay and Umnak Island near Dutch Harbor, Buckner, lacking appropriations and in General Marshall's confidence, diverted funds earmarked for construction on interior army installations to expedite the existence of his two new extension bases of his fortress.

With a handful of obsolete hand-me-down fighters and bombers, the now well-known Alaskan defense commander applied pressure on Washington, D.C. and the Army for newer and bigger bombers and fast, modern fighters. His dry, cryptic wires and messages were answered with promises. With the demand for U. S. aircraft in Europe and elsewhere, priority for Alaska delivery was nil. By late 1941, with deteriorating Japanese relations, the U. S. Government focused its attention to a Pacific powder keg that had all the earmarks of a possible Second World War. A few newer military aircraft started to find their way north to the impatient Buckner.

Elmendorf-based Douglas B-18A bomber of the 73rd Bombardment Squadron (m) taxis to parking area at the Fairbanks' municipal airport following a reconnaissance mission over interior Alaska.

Crosson Collection

Morrison-Knudsen Photo

Reeve and many other bush pilots trained Army flight crews in northern flight techniques and winter aircraft maintenance, in addition to flying "around the clock" airlifting men and supplies. Reeve's Fairchild 71 is pictured above at a remote secret army installation under construction.

Below: Alaska defense forces consisting of 774 enlisted men and 30 officers arrived at Fort Richardson April 27, 1940, under the command of Lt. Col. Earl Landreth. Named in honor of Brig. Gen. Wilds P. Richardson, United States Army, Fort Richardson became an important, permanent military installation.

Elmendorf AFB

Mills Photo

Douglas C-47 with Washington, D.C. fuselage insignia parks at the Fairbanks civilian airport while Western Defense staff officers inspect construction progress of cold-weâther testing station at Ladd Field. First Air Corps detachment assigned to an Alaska station arrived at Fairbanks April 14, 1940.

Right and Lower: Two Boeing B-17 bombers arrived at Ladd Field in November, 1940, for testing and development of cold-weather flight and maintenance operations. In sub-zero temperatures, oil becomes like molasses, rubber turns brittle and breaks, rivets drop out of contracting metal and a solid airplane becomes a maze of loose parts. Pilots were without cabin heat, windshields froze and instruments became inoperative or unreliable. Veteran flyers like Joe Crosson and Bob Reeve worked closely with Major Everett Davis and Captain Dick Freeman adapting winterizing methods worked out by many bush pilots. Both B-17s were reassigned to the 36th Bombardment Squadron at Cold Bay in May, 1942. On June 4, 1942, during the Battle of Dutch Harbor, Lieutenant Thomas F. Mansfield and his crew were lost when one of the early Boeings blew up during a low-level bomb run on the Japanese cruiser "Takao." Captain Freeman was killed later in the crash of the second obsolete bomber.

Elmendorf AFB

Elmendorf AFB

Right: Woodley Airways of Anchorage was one of the many early airlines that flew long hours in close support of the Army, CAA and Morrison-Knudsen construction company. In 1942, Merrill Field had more air traffic than La Guardia Field, New York.

Kellogg Collection

Elmendorf AFB

Left: Temporary nose hangar at Ladd Field served for maintenance out of the 50-degree below weather until permanent facilities were complete. Mechanics soon found that a five minute gloveless operation could cost some fingers and every piece of equipment had to be thawed before use.

Below: Rare photo of early model Curtiss P-40 sans famous airscoop being refueled at Ladd Field.

Elmendorf AFB

Elmendorf AFB

Curtiss P-36 aircraft of the Anchorage-based 18th Pursuit Squadron are lined up in above photo to be refueled at Ladd Field, Fairbanks. On February 20, 1941, 20 P-36s arrived at Anchorage in crates. By August 1941, only nine were operational as a result of crashes, fatigue and lack of spare parts.

Cross Photo

Right: Early Alaska bush pilot, Major John M. Cross, recalled to active duty as a captain in 1941, was assigned to the Ladd Field cold-weather testing station in Fairbanks as an engineering officer. In 1942, he was promoted to major and reassigned to the Eleventh Air Force Service Command at Elmendorf. Cross enlisted in the aviation section of the Army Signal Corps November 1917 as a flying cadet and was commissioned a 2nd Lieutenant pilot-officer in August 1918. After changing to reserve status in 1923, he resumed flying as a civilian and first flew at Cordova, Alaska, in the mid-1930s. Cross flew briefly for Wien Airways at Fairbanks before starting Northern Cross Airways at Kotzebue in 1938.

Mills Photo

Left: Author stops for tire inspection of Alaska Road Commission gravel truck he drove during the summer of 1941 in the construction of the Glenn Highway between Palmer and Glennallen. This vital link between Anchorage and Fairbanks was the first major road building project in the Anchorage area since 1935. The airplane — for decades — was to be the main mode of travel.

Crewmen assist Reeve in loading a 9,000-lb. boiler into specially constructed side opening of a battered Boeing 80-A at Nabesna, Alaska. Reeve, under contract to Morrison-Knudsen, delivered the boiler to a CAA-built air base at Northway. In five months of 1941, he also flew more than 1,100 tons of equipment and 300 men into the site, often carrying three times the rated payload of the tri-motor biplane.

Reeve Photos

Buckner's heavy demand for equipment and supplies for the defense build-up of Alaska was thwarted by the lack of available ships; the only means of transport. The ancient vessels of the Alaska Steamship Company, like the flagship SS "Aleutian," right, sailed the hazardous six-day voyage to Seward loaded to the gunwales with everything from bulldozers to toothpaste.

Rough, icy roads were common at Fort Richardson during the construction period, 1940-41, as they were at all the new military bases. Accidents such as the one pictured at right were not unusual. Added to the bedlam were hundreds of discontented draftees who found themselves assigned to what they considered was the end of the earth. Some deliberately broke laws to be sent to prison, but back to the States.

Elmendorf AFB

Below: Row of parked P-36 pursuit planes can be seen behind huge hangar under construction at Elmendorf Field, Anchorage, in August 1941. The north-south runway was completed June 11, 1941. Elmendorf Field was named in honor of Captain Hugh M. Elmendorf, A. C., who was killed in an airplane accident in the vicinity of Wright Field, Ohio, January 13, 1933.

Elmendorf AFB

Elmendorf AFB

Ladd Field hangar nears completion in May 1941. The Fairbanks air base was designated Ladd Field in December, 1940, in honor of Major Arthur K. Ladd, A. C., who was killed in an airplane accident at Dale, South Carolina, on December 13, 1935.

Reeve Collection

Right: Lt. Col. W. L. Wardell, who commanded the Alaska Communications System of the Army Signal Corps in World War II.

Canadian Archives

Canadian Archives

Battle-proven Royal Canadian Air Force P-40 "Kittyhawks" stationed in western Canada and Yukon Territory were key aircraft in the defense of Canada and Alaska. Right: Groundcrew replaces a tailwheel.

Morrison-Knudsen Archives

Upper: Stinson model "T" trimotor monoplane, under contract to Morrison-Knudsen Company, is refueled from five-gallon cans in preparation to airlifting needed workmen and supplies to an interior airfield construction site. M-K engaged in a broad spread of military construction in Alaska and the company had an air fleet that numbered as high as 17 airplanes. Over half the total dollar volume CAA work was done by M-K.

PBY, while on patrol, flies over Katmai Crater located in the Aleutian Range, 90-miles northwest of Kodiak next to the famous Valley of 10,000 Smokes. On June 6, 1912, Mount Katmai blew its top in the most violent volcanic eruption ever recorded on the North American continent. The explosions were heard as far as Juneau, 750 miles away. The now water-filled crater is three miles across and 3,000 feet deep.

Russell Collection

Canadian Archives

Canadian Archives

Above: Groundcrew servicing Curtiss P-40 fighter aircraft 102-"D" of No. 14 Squadron, RCAF, Umnak, Alaska, June 1943.

Canadian Archives

Right: Pilot Officer A. C. Fanning, left, and Flight Officer Louis Cochand rub down their skis beside the Squadron No. 14 P-40 fighter.

Opposite: Pilots of No. 14 Squadron, RCAF, with Curtiss "Kittyhawk" aircraft in Alaska.

Elmendorf AFB

Above: The day Hitler invaded Poland in September 1939, there was one active military establishment in Alaska. It was manned by 11 officers and 286 enlisted men. This was Chilkoot Barracks near Haines in southeastern Alaska. In this 1941 photo of barracks construction, Fort Richardson, Anchorage, was to become the largest Army base in Alaska before the end of World War II.

Below: Scene on the Ladd Field flight line of North American B-25 bombers in -35-degree temperature.

U. S. Air Force

Elmendorf AFB

Elmendorf-based Lockheed B-34, with General Buckner, is enroute to secret Aleutian bases for inspections and conferences at Cold Bay and Umnak in November, 1941.

Reeve Collection

Right: General Buckner, left, confers with the base commander at Cold Bay. Located at the southwestern tip of the Alaska peninsula, Cold Bay Airfield at Fort Randall later was named Thornbrough Army Airfield in honor of B-26 bomber pilot Captain George W. "Wayne" Thornbrough who was killed during the Battle of Dutch Harbor. Under CAA supervision, Col. Benjamin B. Talley bossed civilians in construction of the field under the guise of building a cannery. Lacking War Department appropriations, Buckner rerouted funds earmarked for new interior bases and created a ficticious Saxton and Company to shroud the building of the Corps' installation.

Jensen Collection

"Aichi" D3A1 (Allied code name VAL) dive bomber of Japanese aircraft carrier "Soryu" is hoisted from Pearl Harbor sans tail assembly. The 1,000 h.p. Mitsubishi-powered Val, one of 15 of its type lost December 7, 1941, participated in the second attack-wave under the command of Lieutenant Commander Takashigi Egusa. Note intact three-bladed Sumitomo-Hamilton propeller.

Jensen Collection

ALASKA'S DARKEST HOUR

Japanese Bomb Dutch Harbor

CHAPTER 3

THE SURPRISE ATTACK on Pearl Harbor by the Japanese startled the world and electrified all Alaska. Within hours, tight security, blackouts and mass military control were instituted and rumors ran rampant. Misread Alaska messages to the "South 48" had already placed Alaska in Japanese conquest.

For the first time since its purchase by the U. S. Government, thousands of Americans were shocked into the realization that Alaska was a part of America and the full awareness of their northern possession's strategic proximity to Asia and Japan.

Buckner's steadfast prognosis of events to come while building the defense of Alaska paid dividends as he swung his forces into high gear. With Elmendorf Air Base as the prime center of operations, the Eleventh Air Force was alerted to move out. The Stateside War Department began to divert a few more bombers to the Northern Command.

At Kodiak in May 1942, nine ships of the newly assigned North Pacific Fleet under the command of Rear Admiral Robert A. Theobald steamed into port. Soon after his arrival aboard his flagship cruiser *Nashville,* Theobald met with Buckner and a clash ensued that prompted Theobald to request from his boss, CINCPAC Commander Admiral Chester W. Nimitz, to clarify the command of Alaska's defense. The issue was sidetracked with a message from Nimitz that intelligence reports placed a Japanese Naval Task Force enroute to the Aleutian Islands.

The Japanese aircraft carriers *Ryujo* and *Junyo,* supported by heavy cruisers *Takao* and *Maya,* three destroyers and an oiler, under the

Pan American Airways' Lockheed Lodestar 18 departs from barrage balloon-encircled Boeing Field, Seattle, on priority flight to Juneau. Overnight-imposed wartime restrictions on flight operations placed a premium on seat space to Alaska.

Gordon Williams Photo

U. S. Navy

The bombed battleship "Pennsylvania" is berthed in drydock behind the heavily damaged destroyers "Cassin," right, and "Dunes," left, shortly after the Pearl Harbor attack. The 26-year-old capital ship fired her main batteries for the first time in an offensive action six months later during "Operation Landcrab," the first U. S. Infantry amphibious landing in history, while reclaiming the island of Attu from the Japanese.

flag of Captain Tadao Kato, steamed through northern fog enroute to the Aleutians. Abeam was Vice Admiral Boshiro Hosogaya's northern force of cruisers, *Nachi, Abukuma, Kiso* and *Tama,* nine destroyers, three troop-laden transports and an escort of several submarines. Kato's orders to hit Dutch Harbor on June 3, 1942, by Imperial Fleet Admiral Yamamoto was a diversionary tactic to screen his strike on Midway Island set for the following day.

Back at Kodiak, Theobald, with the Eleventh Air Force now under his command, disagreed with Buckner that Dutch Harbor was the Japanese prime target and that U. S. forces should move in to protect it. Believing it was risky to assign all his defense at one point, Theobald's strategy was to patrol the Aleutian chain by aircraft and surface vessels and try to spot the enemy invading force. His orders to move the Eleventh Air Force out to advance air bases of Cold Bay and Umnak met stiff opposition by its commander, Brigadier General William O. Butler. Overruled, Theobald moved the fighters and bombers out of Elmendorf to the Aleutian bases. Foul weather deterred PBY pilots and Army bomber crews from conducting an effective search. The Alaska defenses waited.

Air raid sirens screamed across Unalaska Island from Fort Mears to Dutch Harbor when, through the eye of an overhead storm, Japanese attack planes from the deck of *Ryujo* made their strike on still under construction Dutch Harbor, Army's nearby Fort Mears and across the bay village of Unalaska.

Communications failure between Dutch Harbor and nearby fighter-based Umnak broke down, leaving anxious P-40 pilots unaware that the enemy had struck. Fighters scrambled at Cold Bay, 180 miles east of Dutch Harbor, to the plain-English message which said in part, "About to be bombed."

Unopposed, Japanese dive bombers and fighters rallied after twenty minutes of hammering their target and disappeared into the scud heading back to their flattop. Notwithstanding the loss of a PBY shot up while taking off and the victims at Fort Mears, the seemingly devasted Dutch Harbor area had suffered only moderate damage. The Battle of Dutch Harbor had commenced and war had been inflicted on U. S. territory.

Following the first attack, float planes launched from the cruisers *Maya* and *Takao* lost in the storm were suddenly spotted over Umnak Island and P-40s of the 11th Fighter Squadron climbed from their airstrip and engaged the surprised Japanese fighters; minutes later two of them were shot down.

Search planes combed the area to the south and north of the chain. Two PBY Catalinas stumbled on the full Japanese carrier and were pounced on by an umbrella of protective Zeros, shooting down both flying boats, but not before the second had radioed a position report. Army and Navy pilots and crews worked closely trying unsuccessfully to sink the flotilla.

The next day, a second attack was launched against Dutch Harbor from the Japanese carriers, following which the task force vanished into the Aleutian weather.

Russell Collection

May 1942 aerial photo clearly portrays all components of the U. S. Naval Base at Dutch Harbor. At upper left is the hangar and revetment area and in the harbor to the right is the docked SS "Northwestern," which served as a power plant and billeting for civilian construction workers. At lower left is Fort Mears. Located approximately 800 air miles west of Anchorage, Dutch Harbor on Amanak Island had no airfield for land-based protective fighters. The secret army air base, located 40 miles west on Umnak Island, created a situation that former governor of Alaska Ernest Gruening later observed as something like a blindman carrying a lame man on his back; it worked, but not too well. When the Treaty of Naval Limitations of 1922 with Japan not to fortify the Aleutians expired in 1934, the U. S. Navy was quick to resurvey the chain in 1935 for location of possible naval bases. Installations at Kodiak and Dutch Harbor were selected for immediate construction under Federal appropriations approved in early 1940. As late as 1939, Japanese submarines and surface vessels had openly compiled almost a decade of exploratory data of virtually every island in the Aleutians.

Russell Collection

With air raid sirens wailing on June 3, 1942, one of eleven Japanese-type Kate attack bombers from the aircraft carrier "Ryujo" lays a direct hit on an oil storage tank at Dutch Harbor. The Kates, escorted by six Zeros, battled heavy rain and fog to their cloud-covered target. A sudden clearing in the overcast allowed them to complete their mission without U. S. fighter opposition. Six Zeros and 12 Val dive bombers of the air carrier "Junyo" were intercepted by PBY Catalinas; after air battle, they aborted the bombing mission and returned to their flattop.

National Archives

Led by Lieutenant Masayuki Yamaguchi, 15 dive bombers of the Japanese carrier "Ryujo" bombed Dutch Harbor on June 3, 1942. In above photo, center, the ancient Alaska Steamship Company's SS "Northwestern," beached at the harbor dock earlier by a williwaw, takes a direct hit. Fast action saved the vessel.

Below: Japanese bombs cause widespread fires at Fort Mears. One bomb hit an Army barracks killing 25 and wounding many more. Although first-wave bombers escaped, two of four float planes catapulted from the cruisers "Takao" and "Maya" were shot down by the Umnak-based P-40 Warhawks of the Eleventh Fighter Squadron.

National Archives

Russell Collection
SS "Northwestern" burns in upper photo. The Japanese aircraft carried 1,000-pound bombs through heavy flak to reach their targets. The damage to Dutch Harbor was not as devastating as the enemy assumed.

Upper: Badly damaged hospital at the village of Unalaska, across the bay from Dutch Harbor, was bombed during the Japanese attack on June 4, 1942.

Russell Collection

Russell Collection

Right: Stern view of the blazing SS "Northwestern." In 31 years of service with the Alaska Steamship Company, the ancient vessel ran aground 16 times in the treacherous narrows of the inside passage to Alaska off the western coast of Canada.

U. S. Air Force

P-40 Warhawks of the 11th Fighter Squadron under the command of Major Jack Chennault, son of General Claire Chennault of Flying Tiger fame, rose to repel the Japanese Dutch Harbor attack from secret airstrips at Umnak and Cold Bay. Imperial intelligence and scouting led Japanese Task Commander Admiral Kakuji Kakuta to believe the nearest U. S. aircraft were on the Alaska mainland or Kodiak Island. The diversionary engagement at Dutch Harbor cost the Japanese the Battle of Midway — and perhaps the war.

U. S. Air Force

Feisty, dedicated Ernest Gruening, Governor of Alaska, immediately flew to Dutch Harbor to personally inspect the bomb damage. Gruening worked closely with General Buckner and championed many Washington, D.C. concessions to defend Alaska.

Russell Collection

Navy Aerographer's Mate William C. House (top center in dark hood) with mascot, "Explosion," and the Kiska weather team were captured when the Japanese invaded Kiska on June 6, 1942. (Men standing 4th and 6th from left are not members of the team). Explosion was one of several dogs that greeted U. S. troops when Kiska was retaken 14 months later.

National Archives

Russell Collection

First Zero captured intact by the Allies in World War II was this fighter flown by Flight Petty Officer Tadayoshi Koga. While engaged in an exchange of gunfire on June 4, 1942 with a Navy PBY, a lucky bullet from Aviation Machinist's Mate W. H. Rawl's blistergun severed the Zero's oil pressure line. Koga, thinking his engine would pack-up when his oil gauge dropped to zero, picked prearranged Akutan Island, east of Dutch Harbor, for a forced landing. From the air the seemingly flat, clear, emergency field belied the deep bog that tripped the fighter on its back, killing Koga instantly. The valuable prize was sighted six days later by a Navy PBY crew and later a Navy ground team found the Zero only slightly damaged. Dismantled and shipped to the States, the engine was found to be unharmed and soon U. S. test flights were made. Many of the design and engineering features were incorporated into the deadly U. S. Navy F6-F Grumman Hellcat.

Russell Collection

Boeing Co.

Above early model B-17, affectionately named "Old Seventy," was flown by 36th Bomb Sqd. Commander Capt. Russell in a futile attempt to locate the invading Japanese Naval task force before they could strike. Capt. Jack Marks bombed the Japanese cruiser "Takao" on June 4, 1942, the second day of the Battle of Dutch Harbor. On July 17, 1942, Marks and his crew in "Old Seventy" ran out of luck when they flew into a fog-shrouded mountain following a skirmish with Japanese Rufe fighters.

National Archives

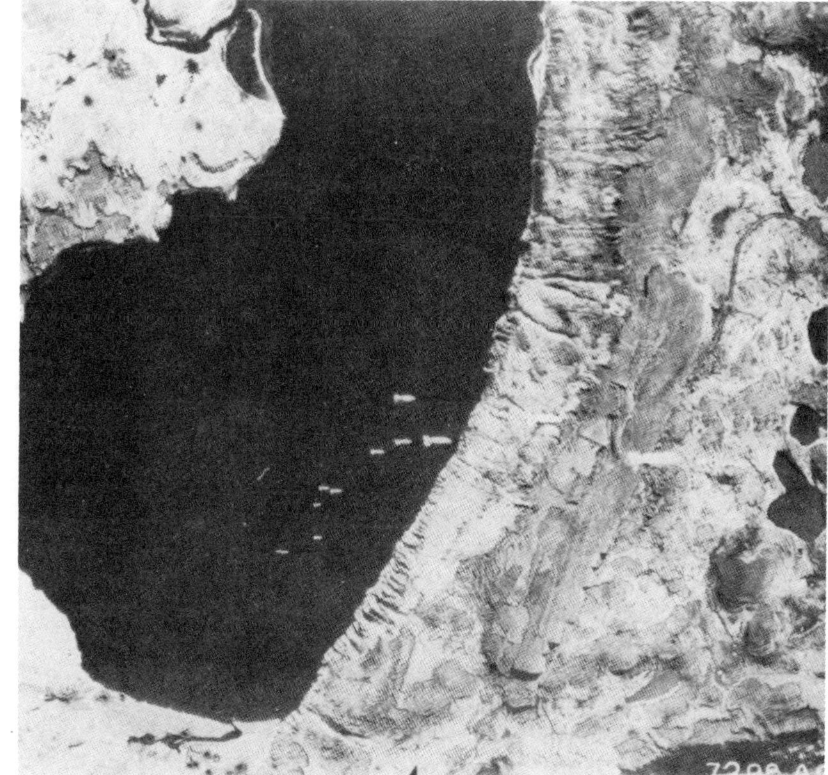

Bombs fall on Japanese-held Kiska during the early hours of their occupation. In 1904, Kiska was set aside as a naval reservation. The large island's huge harbor offered sheltered anchorage, although ground swells were not uncommon in the otherwise calm waters.

Elmendorf AFB

Eleventh Air Force bomber commander Colonel William Olmstead Eareckson led a handful of obsolete bombers of the 36th Squadron in search of the elusive Japanese Naval task force following their attack on Dutch Harbor. Without food or sleep, his pilots and crews joined Navy PBYs in an effort to find and sink the enemy. Born in Baltimore, Maryland on May 30, 1900, Eareckson enlisted as an army private in World War I and subsequently was appointed to West Point. Following his commission, he devoted his entire military career to military aviation. Assigned to the defènse of Alaska as a major in March 1941, the dynamic, headstrong, crack-pilot quickly assumed vigorous leadership of the Aleutian-based 11th.

Russell Collection

Kodiak-based Consolidated PBY "Catalina" is parked in freshly excavated revetment area. Navy Patrol Squadrons VP-41, VP-42, and VP-43 all served in the Aleutians during World War II.

Above: Seven brand new Boeing B-17E Flying Fortresses arrived at Cold Bay on June 7, 1942, supplying Col. Eareckson with the aircraft he needed to lead the first bombing mission on the Japanese invaders of Kiska and Attu. The B-17 was operational all through the Aleutian campaign, although the Consolidated B-24 replaced it as the Eleventh Air Force basic bomber.

Boeing Co.

U. S. Air Force

The 21st Bombardment Sqd., assigned to two weeks routine patrol duty, arrived at Cold Bay in the heat of Eareckson's blitz of Kiska. Within hours, their bombers were fueled, bomb-loaded and on their way to bomb the Japanese-held island.

Caption translation of Japanese photo above reads: "12 June 1942, at about 09:10. Enemy PBY flying boat receiving intense a/a fire. Immediately after this it dropped in altitude, trailed smoke and fled. It is considered to have made a forced landing. The ship at left is the 'Awath Maru'." Puff of smoke on Kiska mountainside to the left of PBY is crash of B-24 piloted by Capt. Jack Todd, new commander of the 36th Bomb Squadron, who with Navy Lt. Clark Hood, as navigator, caught a direct flak hit in their open bomb bay just after their bomb release.

Right: Armourers attaching Victory Loan pennant to Kiska-bound bomb in front of Curtiss "Kittyhawk" I aircraft AK987 of No. 14 Squadron, RCAF, Umnak, Alaska.

Flak-damaged B-17 struggles to stay airborne as it returns to Umnak after a bombing raid on Japanese ships in Kiska Harbor. Heavily entrenched enemy antiaircraft guns and warship armament forced U. S. Army and Navy bombers to fly single sorties at low altitudes in an effort to surprise the invaders.

After the seven-day Kiska Blitz, the Japanese had sustained negligible damage and the Americans suffered a retreat. The western 1,000 miles of the Aleutians was "no-man's land" and Japanese-held.

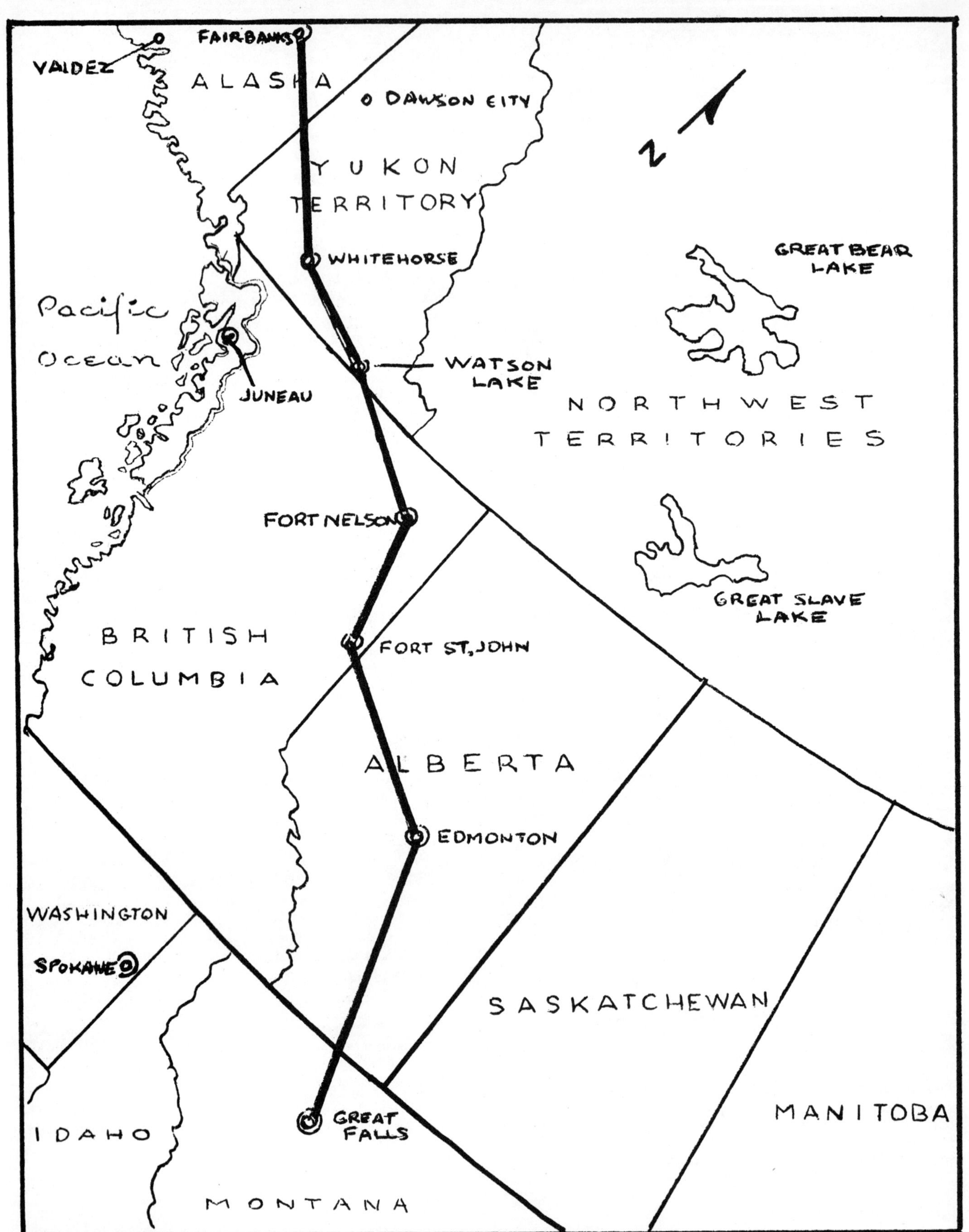

FAIRBANKS
VALDEZ
ALASKA
DAWSON CITY
N
YUKON
TERRITORY
WHITEHORSE
GREAT BEAR
LAKE
Pacific
Ocean
JUNEAU
WATSON
LAKE
NORTHWEST
TERRITORIES
FORT NELSON
GREAT SLAVE
LAKE
BRITISH
COLUMBIA
FORT ST, JOHN
ALBERTA
EDMONTON
WASHINGTON
SPOKANE
SASKATCHEWAN
GREAT
FALLS
MANITOBA
IDAHO
MONTANA

GRAVEYARD OF AIRCRAFT

Million Dollar Valley of Crashes

CHAPTER 4

THE EARLY DAYS OF Alaska's defense build-up required the supply of as many military aircraft that could be assigned to the Northern theater of operations. At times, green pilots flying fast, powerful and sometimes unforgiving airplanes over the desolate, rough terrain in adverse weather took its toll in lives and badly needed bombers and fighters. Whole squadrons were known to have become separated and forced down all along the route in northern Canada. On one occasion, two squadrons of B-26 Mauraders headed north from California and one month later, the first plane arrived at Fairbanks. Many of the planes crashed in a mountain pocket northwest of Watson Lake, Yukon Territory, that became known as "Million Dollar Valley." When the last of the twin-engine bombers arrived at Fairbanks, 45 days since the start north, a total of 13 had failed to make it.

Later, seasoned pilots and veterans of the Air Transport Command (ATC) assumed the responsibility of delivering the valuable airplanes.

With General Gaffney commanding the Northern Division of the ATC with headquarters in Fairbanks, Lend-Lease ferrying of aircraft to Russia was added to the route beginning in September 1942. The sight of Russian personnel in Fairbanks and Nome was commonplace and almost 8,000 aircraft of various types were delivered by ATC to the Russians at Ladd Field, who then ferried them to Siberia via Nome.

The route used by the Army's ATC began at Great Falls, Montana, where most of the aircraft were winterized; then to Edmonton, Fort St. John, Fort Nelson, Watson Lake and Whitehorse, all major stops on the way to Fairbanks. This was a 1,900-mile trip all the way, in comparison with 13,000-mile middle east route around Africa, up the Persian Gulf and across Iran.

Boeing B-17E in sub-zero weather at Ladd Field just after having flown in from Great Falls, Montana, where it had been winterized.

Elmendorf AFB

U. S. Air Force

Lack of navigational aids, foul weather and inexperienced pilots contributed to severe losses of desperately needed fighters and bombers being ferried to Alaska. Thirteen B-26 Marauders of the 77th Bombardment Squadron, enroute from Sacramento to their Northern assignment, resulted in one crash in Canada and four crashes between Edmonton and Fairbanks that, thereafter, became known as the Million Dollar Valley.

U. S. Air Force

Elmendorf AFB

Above: Aerial view of Ladd Field, Fairbanks, where over 8,000 aircraft passed through enroute from States to delivery in the Aleutians or to the Russians.

Jensen Collection

Right: Army Stinson L-5 liaison plane force lands, barely missing above wires.

Skoric Photo

Left: Russian-born Joe Skoric, corporal U.S. Army, served as a control tower operator at Ladd Field during World War II. Serving as a Russian sergeant in World War I, Skoric escaped during the Soviet Revolution and came to Alaska. He learned to fly in Seattle; bought a plane and took it to Nome, where it was wrecked. He served as interpreter for Russian pilots flying in and out of Fairbanks.

Colonel Everett S. Davis, first commander of the Eleventh Air Force, was killed November 28, 1942, when his C-53 Transport aircraft struck a mountainside one-half hour flying time out of Naknek. Below: Three B-25 bombers fly in tribute to Davis over his burial location.

Alaska Army Air Transport Commander Brig. Gen. Dale V. Gaffney, stationed at Fairbanks, supervised the Alaska-Siberia aircraft Lend-Lease delivery program. Many U. S. airline pilots, with Army ferry crews, delivered 7,930 aircraft of various types to the Russians at Ladd Field. When one Russian pilot was missing on a flight between Fairbanks and Nome, his commanding officer declined Gaffney's offer to institute a search saying he wasn't a very good pilot anyway.

U. S. Air Force

Below: Nome-based Soviet air personnel and U. S. Air Corps liaison officer, fourth from right, stand by as a Russian flight crew prepares to ferry American-built B-25 Mitchell bomber to their homeland. Details of the Lend-Lease air operation of the Northern route were a closely guarded secret outside of those army agencies directly involved. Most Russian flyers were fearless aviators and piloted their aircraft as though they were airborne bulldozers.

Reeve Collection

Boeing Co.

Above: B-17 Fortress enroute to Alaska and the Aleutians. Bush pilots Bob Ellis, Kenny Neese, Bert Ruoff, Murrell Sasseen and Clayton Scott were some of the veteran Alaska pilots that ferried aircraft during World War II. Another well-known bush pilot, Art Woodley, was selected by the Alaska Defense Command to fly intelligence service personnel on confidential missions during World War II.

Right: Early Alaska bush pilot Murrell Sasseen and wife Dorothy Sasseen. Both served in the war effort; "Sass" ferried aircraft of all types to Alaska as an Air Force captain and Dorothy winterized aircraft at Great Falls, Montana.

Below: Airfield at Nome. The far north Alaskan city, famous since Gold Rush Days, was the scene of many Russians ferrying aircraft to Siberia.

Saseen Photo

Wien Consolidated

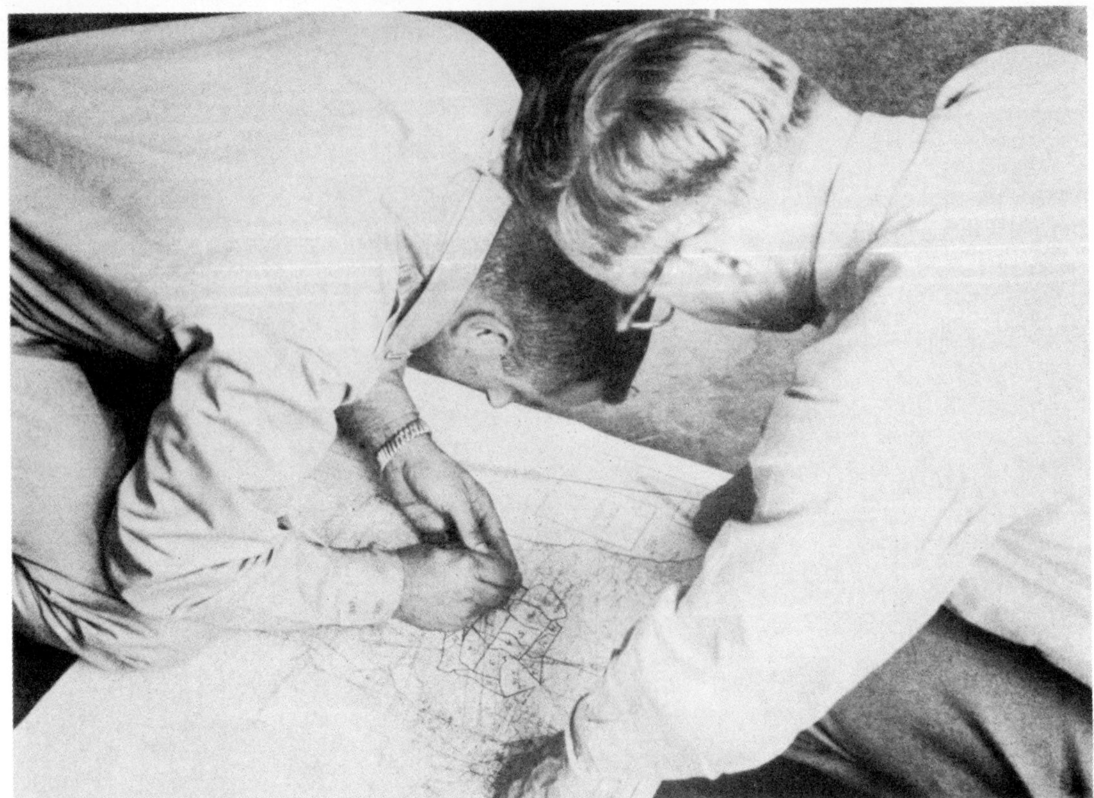

Above: Lieutenant and assistant coordinator plot probable locations of downed aircraft.

U. S. Air Force

Below: Barren countryside where several pilots found themselves when weather closed in.

U. S. Air Force

In 1939 photo at right, Army Sikorsky OA-10 crew members, Lt. Col. Thomas Handy, left, 1st Lt. Harold Smith, center, and West Point cadet Donald Lee, Jr., pause for picture with Helen Welch, left, and Lou Horning of Anchorage during a rest stop at Merrill Field while conducting a survey mission in Alaska. Allan Horning, former military pilot prior to flying the bush out of Anchorage, was ordered to active duty as guide pilot for expedition whose task was to select prospective army air base sites throughout Alaska and the Aleutians. The Elmendorf Field location at Anchorage was one of those chosen during this tour. Horning later joined the Civil Aeronautics Administration (CAA). Prior to and during World War II, the CAA was instrumental in the location, planning and construction of airfields throughout all of Alaska in conjunction with installation of communications, navigation aids and other phases promoting safety and regulating air traffic in the territory.

Horning Collection

Below: (L-R) Lou Horning, Jack Jefford, Al Horning (standing in cockpit), Bill Hanson (partly hidden), Jim Hurst and Larry Lawton are shown here with Douglas amphibian at Skilak Lake in 1943. Except Lou Horning (wife of Al Horning), all were employees of CAA. Al Horning ferried the administration's flying boat from Glendale, California to Alaska. Early CAA pilots flying weather bureau flights were paid $2.50 per hour if they flew to 15,000 feet; otherwise they received no pay.

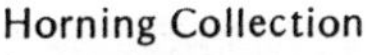

Horning Collection

Russell Collection

Above: May 1942 air view of the east side of huge, mountainous Kodiak Island shows North Pacific Fleet naval base headquarters during the Aleutian campaign, with adjacent airfield to the right. Home of the world-famous Kodiak bear, the island was first inhabited by the Russians in 1791 with the establishment of a trading post and dock facilities for king crab fishermen. It was here that the original clash between Navy's Theobald and Army's Buckner belied the grass roots cooperation and comradery between combat sailor and soldier.

U.S. NAVY AND SQUEAKY ANDERSON

From Catalinas to Yippie Boats

CHAPTER 5

THE U.S. NAVY'S INVOLVEMENT in Alaska dates back to just soon after the Territory's purchase by the United States. On April 3, 1879, the *USS Alaska* arrived at Sitka to establish naval rule as a governing force; on June 14, 1879, the *USS Jamestown* arrived to relieve the *USS Alaska* as headquarters of the Alaska government. The *USS Wachusett* relieved the *USS Jamestown* July 30, 1881.

In 1883, the wooden three-masted Barkentine icebreaker *Bear* joined the *Thetis* and *Alert* in search for the Greely Expedition, missing near Ellesmere Island in the Arctic. The *Bear* successfully found and rescued the survivors on June 22, 1884.

For years Sitka was the naval headquarters of Alaska, with its duties and size increasing throughout the 1920s and 1930s commensurate with the growth of Alaska.

Pacific fleet problems, survey and testing missions were conducted in all Alaskan waters, including the Aleutian chain. Patrol flights were operated out of Seattle periodically throughout the 1930s and intensified naval activities and building in northern Alaska began shortly before 1940, culminating with bases constructed at Kodiak Island and Dutch Harbor.

In World War II, the airplane was an effective and many times decisive weapon. While the aircraft were important in the Aleutians, much must be said for the other naval sections that operated in the world's worst weather and treacherous waters.

The highly skilled ever-present Seabee batallions functioned with precise timing when needed. Actions of picket boats, minesweepers, motor torpedo boats, together with submarines, destroyers, cruisers, battleships and aircraft carriers were necessary components to accomplish the task.

The Navy and Army in the Aleutian war were a close working team despite post-war higher headquarter reports to the contrary. Navy pilots flew with Army crews; Navy crewmen loaded torpedos on Air Corps bombers, and Navy and Army billited together. Sailor and soldier bundled in Army heavy weather gear became uniformed twins and with rank covered, officers and enlisted men looked alike.

A difficult war fought in a desolate forbidding area of the world – the Aleutians – required the coordination of the Army, Navy, Coast Guard; air, land and sea.

James Hartley Photo

In 1935, the USS "Chicago," docked at Seward in photo at right, participated in the U. S. Navy Problem XVI, which included a sweep out the Aleutian chain under the watchful eyes of the Japanese aboard the "fishing vessel" "Hakuyo Maru."

U. S. Navy

Above: Crusty Rear Admiral Robert A. "Fuzzy" Theobald, left, and Lieutenant Charles E. "Cy" Perkins at Umnak Island November 7, 1942. Theobald, North Pacific Force Commander headquartered at Kodiak, ranked high in his 1907 Naval Academy class. Perkins earned the Navy Cross for his single-plane attack on the Japanese Naval task force during the Battle of Dutch Harbor. On one engine, the lieutenant nursed his badly damaged PBY safely back to their home base. Lower: PBY of Squadron 42, commanded by Commander James S. Russell, is beached at Kodiak following a patrol flight out the Aleutian chain.

U. S. Navy

U. S. Navy

Navy Lieutenant Anderson, left, and hastily recalled Alaska cannery owner Captain Carl E. "Squeaky" Anderson, hold a conference at Dutch Harbor, in photo above, taken after the Japanese invasion of Kiska and Attu. "Squeaky," a Swede whose nickname described his loud, piercing voice, was said to be the only sailor alive who really knew Aleutian waters. After training crews for his famous "Yippie" boats (converted commercial fishing vessels for patrol and rescue duty), the colorful Alaskan became a beach-boss of wide legend throughout the Pacific in World War II following his outstanding accomplishments as beachmaster in the treacherous invasion of Attu. The World War I veteran rose to the rank of Rear Admiral in World War II.

Left: PBY blitz pilot William Theis with Patrol Wing Four Commander Capt. Leslie E. Gehres. Theis devised a dead-reckoning bombing pattern used against the Japanese on cloud-covered Kiska. Former carrier-deck pilot Gehres, although not proficient in PBYs, earned the title, "Custer of the Aleutians," in his drive to keep his flying boats in the air against the Japanese.

U. S. Navy Photo

Lower: Kodiak crews thaw out PBY Catalina before starting engines. Soon after the Battle of Dutch Harbor, 20 new PBY-5As of Patrol Squadron VP-43, under the command of Lt. Commander Carrol B. "Doc" Jones, arrived in the Aleutians to back up Commander Russell's P-42 Squadron which was now reduced to 14 planes from the original 23. Both squadrons participated in the Kiska Blitz, nicknamed "PBY Elimination Center" by crewmen, proving the Catalina's combat capabilities.

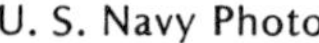

U. S. Navy Photo

U. S. Navy Photo

Seaplane tenders, as floating air bases, met all the needs of the PBYs and their crews. They made air-striking forces deep in the Aleutian chain possible. The tender "Casco," in earlier photo above, at her usual Nazan Bay, Atka Island anchorage, was torpedoed in August 1942 by the Japanese submarine RO-61, killing five men and injuring 20. Two PBYs, piloted by Lieutenants Samuel Coleman and C. H. "Bon" Amme, ruptured the RO-61's hull with their bombs and the *coup de grace* came from the depth charges of the destroyer "Reid." Other tenders that served in the Aleutian Campaign were the "Gillis" and "Williamson." The Navy's fleet air wing of PBYs and PVI Venturas flew 704 combat sorties and uncounted thousands of patrols, dropped 590,000 pounds of bombs at a cost of 52 aircraft, including 16 in combat. Some of the surviving PBYs are still flying today.

The effect of violent 100-mph Aleutian winds is depicted in photo at right. The saying was that if the wind ever stopped, everybody would fall on their face! The winter of 1942-43, one of the worst in Aleutian history, added to the misery of friend and foe alike.

Reeve Collection

Lower: Wrecked PBY caught in a williwaw and flipped on its back.

U. S. Navy

Russell Collection

Photo above is VP-42 Squadron PBY flying boat on Aleutian patrol. As a result of their constant sweep of the chain, a Navy pilot turned in the first contact report of Admiral Kakuta's task force penetration of Aleutian waters on June 2, 1942. Originally built in 1939, the slow, lumbering Catalinas earned meritorious recognition as an effective offensive aircraft throughout the Aleutian campaign.

Russell Collection

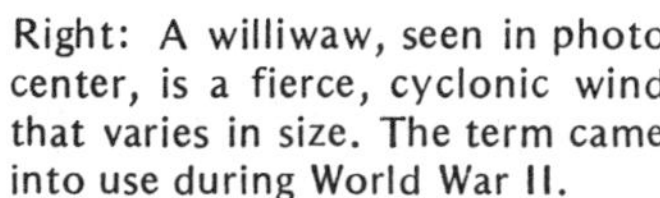

Right: A williwaw, seen in photo center, is a fierce, cyclonic wind that varies in size. The term came into use during World War II.

Russell Collection

1942 air view shows Dutch Harbor's prominent Mount Ballyhoo protruding up from Unalaska Bay in center of upper photo. In 1902, 20 acres was set aside by the U. S. Government at Dutch Harbor for a coal refueling station. The following year proposals were made to have it fortified. A radio station was opened in 1912 for commercial messages.

Russell Collection

Above: Skipper of Navy PBY Sqd. 42, Lt. Commander James S. Russell, left, popular movie comedian Joe E. Brown, and Kodiak Army Base Commander Brig. Gen. Charles H. Corlett, right, pause during tour of base.

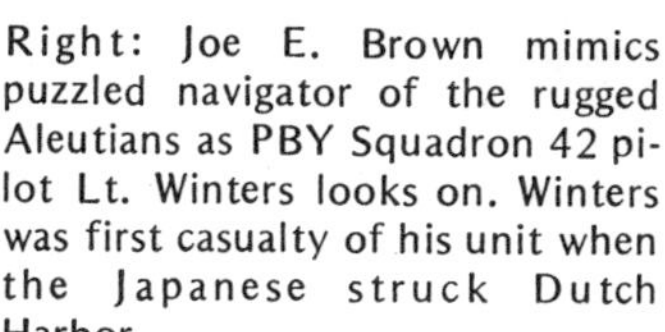

Right: Joe E. Brown mimics puzzled navigator of the rugged Aleutians as PBY Squadron 42 pilot Lt. Winters looks on. Winters was first casualty of his unit when the Japanese struck Dutch Harbor.

Boeing Co.

Boeing B-17s joined B-24s in continued bombing of Kiska on orders to "Get Kiska back." The 36th and 21st Bombardment Squadrons employed every feasible bombing tactic to knock out the Japanese-held fortress. The 1200-mile round trip to Kiska from the nearest bomber base at Umnak, in ice, wind, fog and extreme turbulence, sometimes took up to 12 hours to complete. The Japanese-held Attu, 150 miles beyond, was out of U. S. Army bombing range and closer air bases were needed.

OPERATION LANDCRAB

The Invasion of Kiska And Attu

CHAPTER 6

A GENERAL PLAN for the invasion of Kiska, the largest of the two Japanese-held Aleutian Islands, was submitted to Western Defense Commander Lt. General John L. DeWitt and Chief of the Pacific Fleet Admiral Chester W. Nimitz by North Pacific Fleet Commander Admiral Kinkaid. In January 1943, the plan, which called for an amphibious landing, was forwarded to higher headquarters with the approval of DeWitt and Nimitz.

Lt. General Buckner's seasoned Fourth Army was first selected for the assault, but the U. S. War Department chose desert trained 7th Division commanded by Major General Albert E. Brown.

When the plan had been accepted, Kinkaid supplemented it with a suggestion to bypass Kiska and strike Attu first; thereby, hopefully cutting off Kiska – a tactic that later was adopted in the island-hopping Allied conquest of the South Pacific.

With the original date for the invasion of Attu (code name JACKBOOT) set only three months hence, a series of planning meetings ensued with the heads of all the services involved. The entire operation challenged the wisdom of the most brilliant military strategist; treacherous amphibious conditions, logistic problems of Attu's rugged terrain, coordination and timing of air, land and sea forces, and weather – the one factor in the Aleutians which was notoriously unpredictable.

In below freezing weather May 11, 1943, 244 crack troops of a scout force led by Army's Captain William H. Willoughby silently landed on the beach of Attu from rubber boats, thereby launching the bloody battle of Attu. The Japanese forces on Attu, informed of the impending Allied invasion, were prepared and waiting.

Attu was reclaimed 18 bitterly fought days later. The price had been high with almost 2,400 Japanese dead; 500 by mass suicide. Occupation forces suffered 3,829 casualties; 549 killed, 1,148 wounded, 2,132 victims of frostbite, trench foot and accidents.

Aside from Iwo Jima, it was to become the most costly battle of the Pacific Theater of World War II.

Planning the Kiska Invasion: Seated, left to right: Rockwell, Kinkaid, Corlett, Buckner, Butler, Pearkes (Canada). Behind each officer stands his Chief of Staff.

National Archives

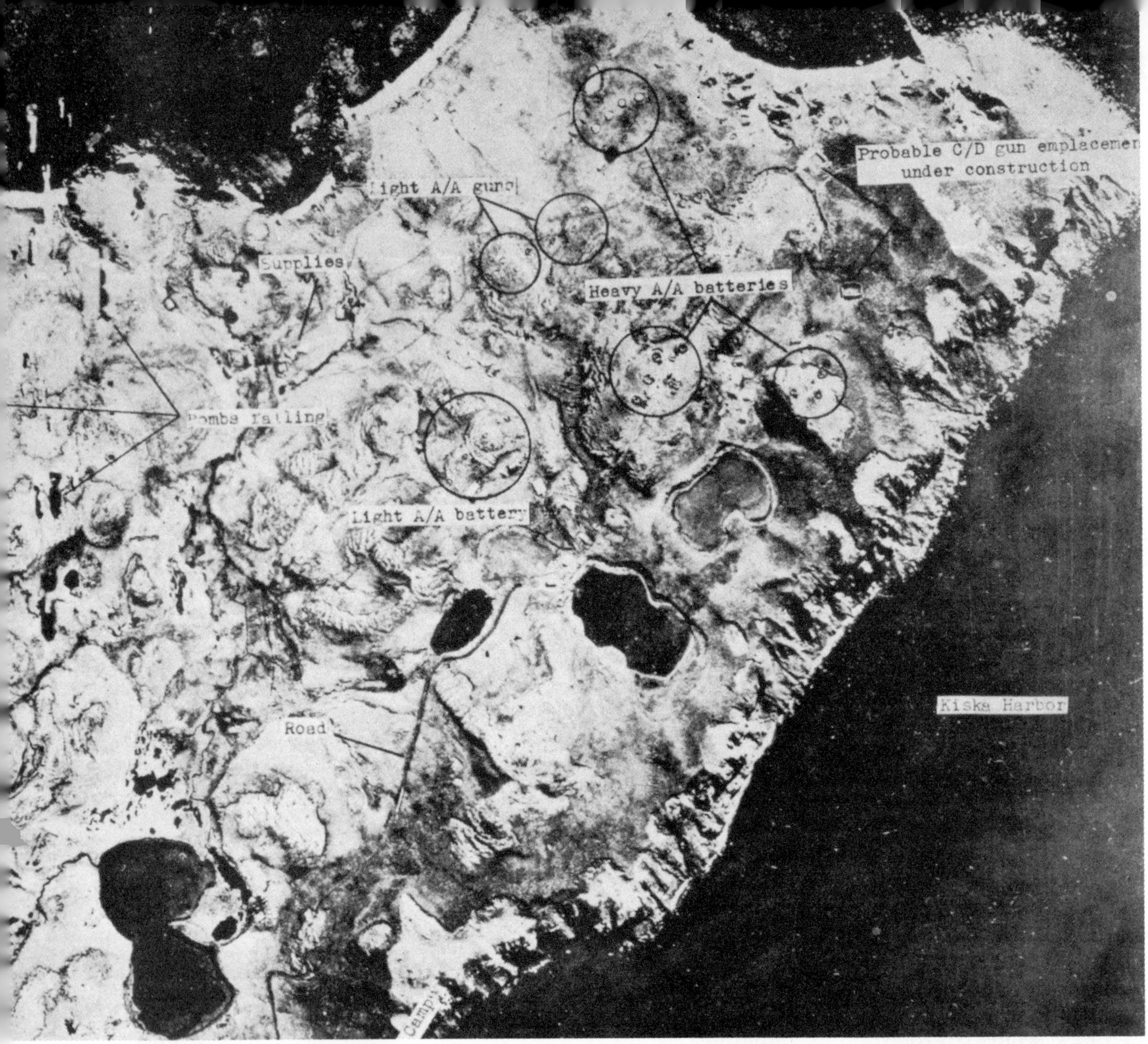

Large scale vertical photo showing Japanese installations of Kiska's North Head. Note how clearly guns can be seen from air.

Photo of Japanese seaplane ramp and building on Kiska was taken from Eleventh Air Force bomber during deck-level attack.

U. S. Air Force Photos

Upper: Heavily laden ships with Eleventh Air Force supplies and equipment is hastily unloaded for use in the blitz of Kiska.

U. S. Air Force

Lower: Vital supplies are hauled along a mountainous road by tractor train to a nearby secret fighter air base.

U. S. Air Force

Left: Fuel and bomb dumps are dispersed over large area at Eleventh Air Corps base.

Below: Sled is used to deliver 100-pound bombs for loading into B-24 "D" model bomber. This plane was one of the "pink elephants," so painted for desert operations, that were rerouted to Alaska by the Air Corps in answer to General Buckner's plea for more aircraft.

U. S. Air Force Photos

Left: Ordinance crew attach fins to bomb in preparation for loading into bomb bay.

Upper: Consolidated B-24 lifts off the runway at Umnak Island for raid on Kiska. The Japanese occupation of Kiska was discovered by Captain Robert E. "Pappy" Speer, scouting in a LB-30 (British version of B-24). Speer and other Army and Navy flyers had searched for days for the Dutch Harbor attackers. New B-24s of the fresh-from-the-States 21st Bombardment joined Colonel William Eareckson's weakened 36th Bombardment in the futile attempt to blast the invaders out of Kiska. Heavy losses of Army and Navy aircraft to Japanese 75mm antiaircraft and the weather soon turned the campaign to one of harassment and concentration on attacking replacement and supply ships. Six months of steady bombing in good and bad weather effected a high toll of Japanese aircraft, supplies and morale. Japan was feeling the cost of trying to hold its remote conquest.

U. S. Air Force Photos

Below: Combined air-naval operations attack Kiska. This panorama taken during mission shows four U.S. ships cruising about 4,000 yards off the island's south shore. The four cruisers shown are: (L-R) "Santa Fe," "Louisville," "San Francisco" and "Witchita." Smoke from funnels above ships is caused by suddenly reducing speed as they maneuvered into firing position. Notice antiaircraft burst at right over little Kiska direct at B-24s paying a simultaneous visit.

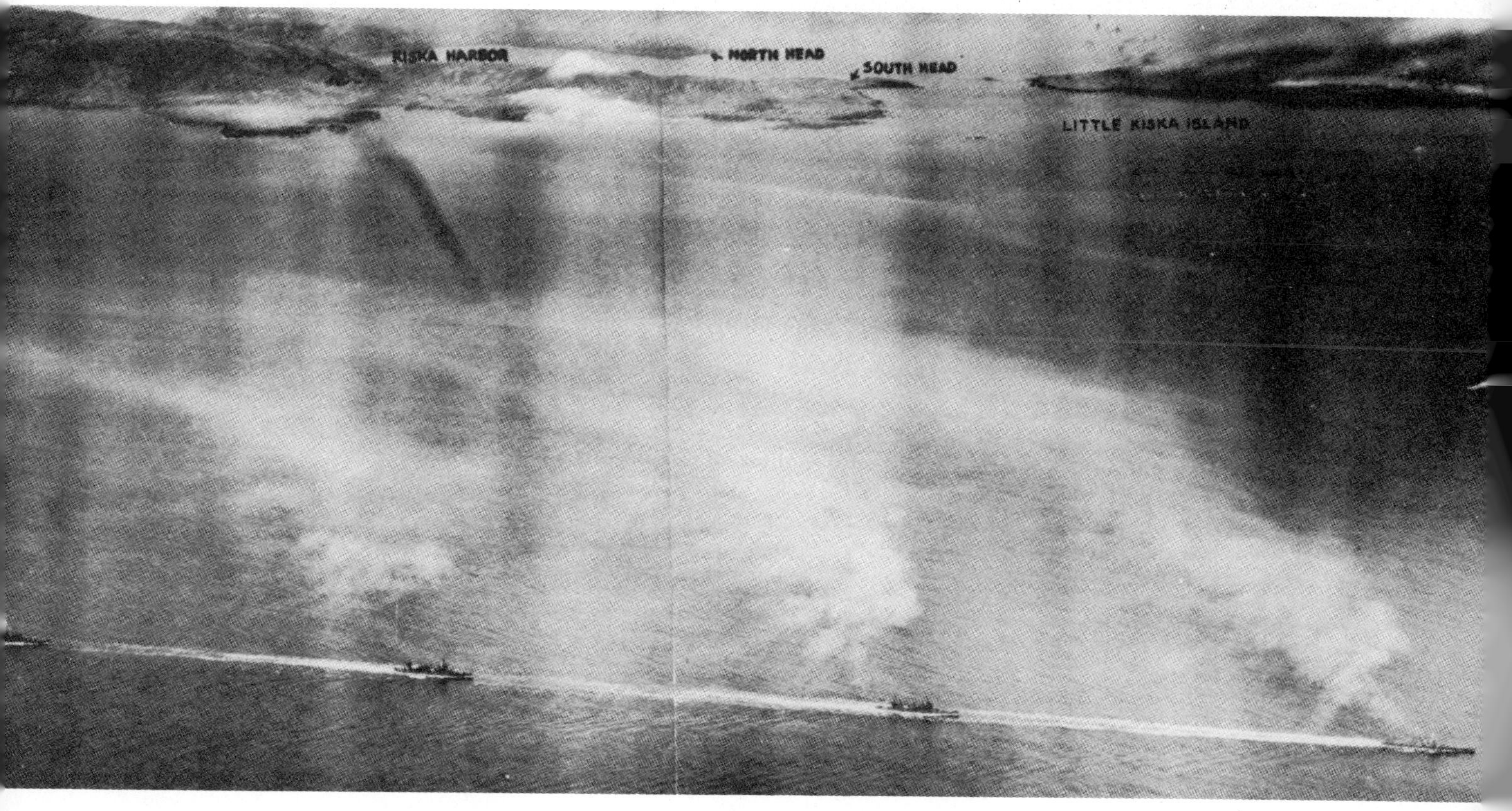

Left: In an attempt to build an airfield, the tremendous number of wasted Japanese man-hours of labor is evident in this picture of the northeast end of Kiska. In left foreground is a completed revetment. Tracks for dump cars used in the construction of the airstrip are clearly visible. Bomb bursts dot either side of the road leading to the main Japanese camp. Intensified U. S. bombing precluded the Japanese completion of an airstrip on Kiska.

Lower: During snowstorm, 36th Bombardment Squadron ground crewmen are quick to assist B-24 pilots to park in revetment area following raid on Kiska.

U. S. Air Force Photos

Boeing Co.

Boeing B-17 Flying Fortresses of the 11th Air Force, predecessor of the Alaskan Air Command, that played an important role in freeing the Aleutian Islands of Japanese invaders during World War II.

U. S. Air Force

Five radar-equipped Boeing B-17s reversed accepted tactics and escorted Lockheed P-38 "Peashooters" far out in the chain to Atka, where Paramushiro-based Mavis bombers were attacking. Fou! weather conditions and the long range prevented the 54th Fighter Sqd. pilots time to pinpoint the enemy raiders. The Col. Eareckson brainstorm paid dividends. The "mickey men" picked up the incoming enemy flight and the 54th split the surprised foe and shot down two of the flying boats. At right is the victorious peel-off at home base.

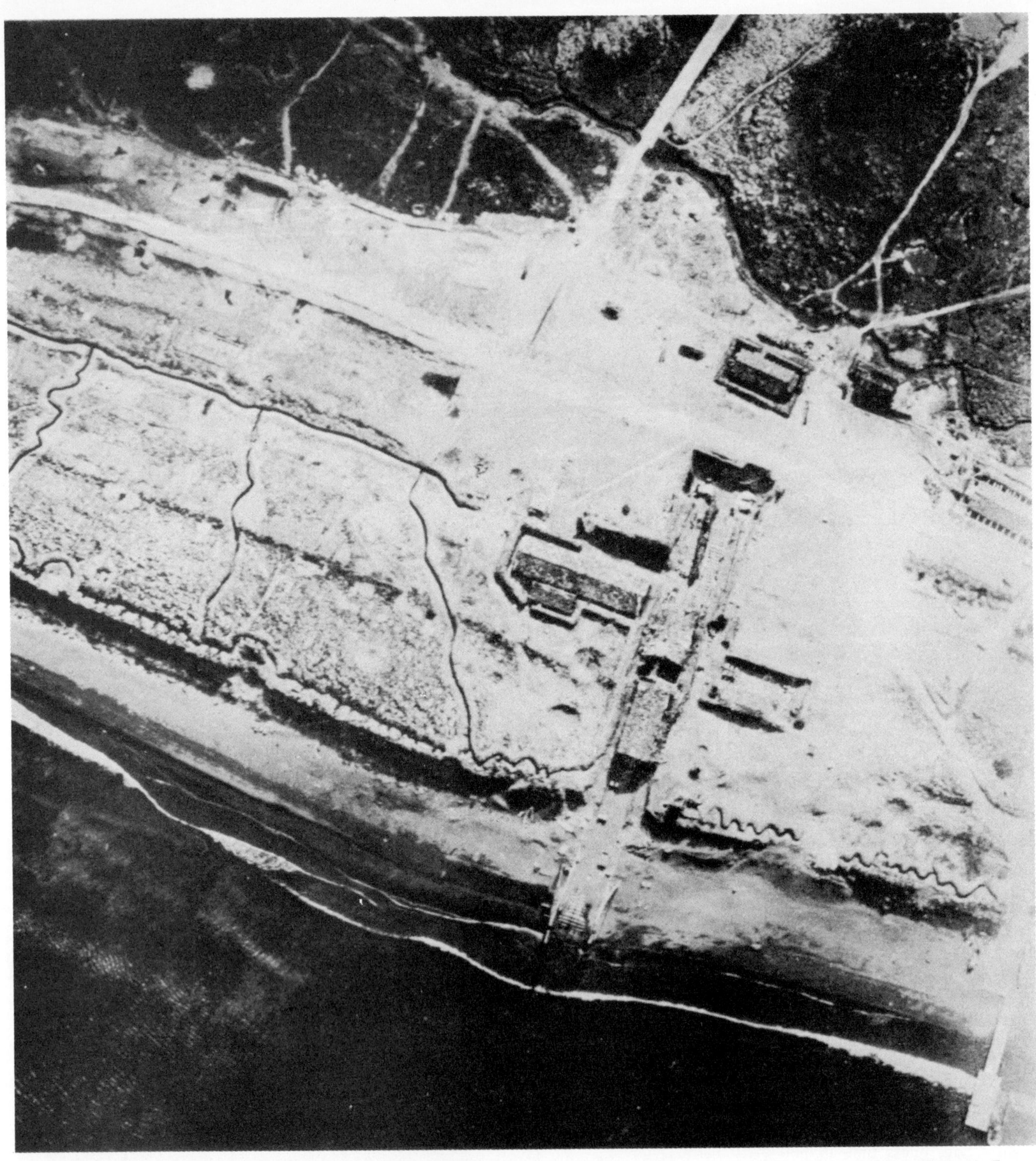

U. S. Air Force

Precision bombing at lower levels, but higher risks, paid dividends. The Japanese built huge revetments but it was futile. This photo shows a submarine base on Kiska Harbor that has suffered extensive damage by Army and Navy constant bombing. One damaged midget sub is still on the ways. The wavy, black lines along the beach are trenches. The Japanese troop strength of Kiska reached about 8,000; near 1,000 on Attu.

National Archives

Upper: Battleships "Nevada" and "Pennsylvania," escorted by the new CVE aircraft carrier "Nassau," join the largest Naval assault task force since Guadalcanal to launch the bloody battle of Attu in May 1943. In a previous 30-day softening-up purge, Brig. Gen. William O. Butler ordered everything that would fly in his Eleventh Air Force to bomb Attu.

Lower Japanese photo is of operations room on Attu prior to the U. S. invasion in May 1943. The troop strength of Attu had been underestimated and that of Kiska overestimated. Attu was selected to be retaken first on the assumption of lesser resistance. The ensuing conflict cost 3,000 American casualties and is recorded as one of the most traumatic actions of World War II.

Reeve Collection

Canadian Archives

Pilots with Curtiss "Kittyhawk" aircraft "P" of No. 14 Squadron, RCAF, Umnak, Alaska, June 12, 1943. (Left to right at front): F/L Al Grimmins, P/O A. C. Fanning, F/O Bill Mac Lean, P/Os Kelling Barrie, Ronnie Cox, F/O George Stiles. (Left to right at rear): F/O Frank Galbraith, F/Sgts. H. Hobbie, Ray Bell. In May 1942, Royal Canadian Air Force responded to the Japanese invasion of the Aleutians by placing several combat squadrons at the North Pacific Fleet Commander's disposal. The RCAF 115th Fighter Squadron was already stationed at Annette Island near Juneau. The 111th war-hardened pilots of the Battle of Britain flew in their Curtiss P-40s eager for combat. Some RCAF squadrons dispersed to various coastal bases as back-up forces for other Allied units that had moved up to the Aleutians.

U. S. Navy

Assistant Secretary of War John J. McCloy arrives at Adak August 11, 1943, four days before the invasion of Kiska, and is met by Western Defense Commander Lt. Gen. John De-Witt. Admiral Thomas C. Kinkaid, champion of six major Pacific battles and replacement for Admiral Theobald, is seen over McCloy's right shoulder. Gen. Buckner joins the group at far right.

Above: Battle veteran P-40 Warhawks are parked at the Alert line ready to join in the invasion of Kiska. Other American fighters flown in the Aleutian Campaign were P-38 "Lightning," P-39 "Aircobra," P-51 "Mustang," and P-82 "Twin Mustang."

7,300 U. S. combat troops invading Kiska on August 15, 1943, were greeted by a half-dozen dogs left hehind when on July 28, 1943, a Japanese task force evacuated 5,200 troops from Kiska in fifty-five minutes and slipped away back to Japan, undetected. Some of the evidence that the enemy had occupied the island was reflected in the wreckage area shown below. The Allied Air Services lost 471 airplanes in the Aleutians during the campaign. The Japanese losses in aircraft were 69 in combat and 200 in fog and storms.

Elmendorf AFB Photos

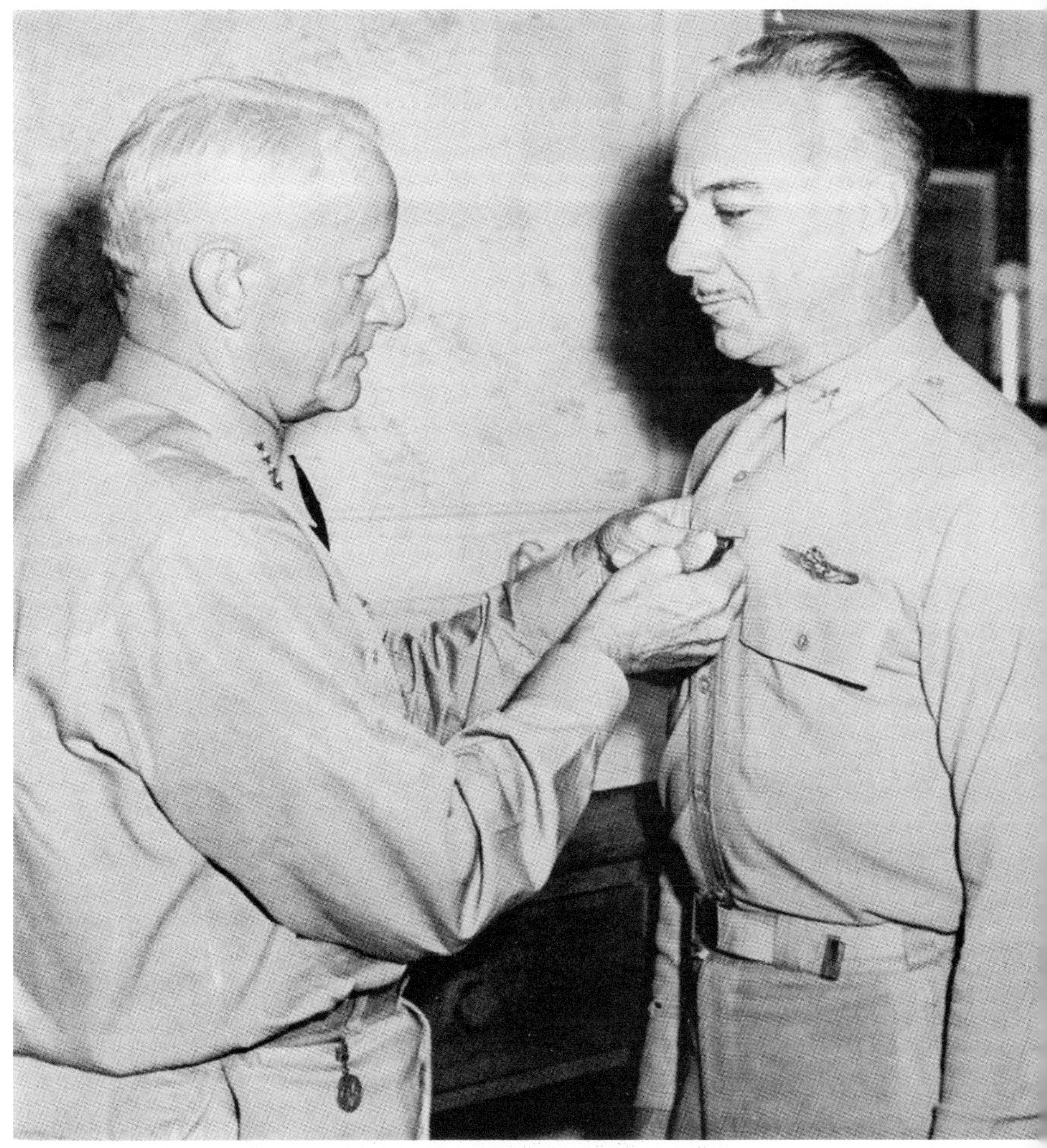

U. S. Navy

Fleet Admiral Chester W. Nimitz pins the Navy Cross on Col. William Eareckson for outstanding tour in the Aleutians. The highly decorated Eareckson served in the Pacific as an air staff officer with the Navy. His outspoken manner and disregard for military bureaucracy affected his promotion. When he retired from the Air Force in 1954, he had been a colonel 13 years. He died October 26, 1966.

Reeve Photo

REEVE SUPPLIES THE ALEUTIANS

Glacier Pilot Becomes ACS Air Force

CHAPTER 7

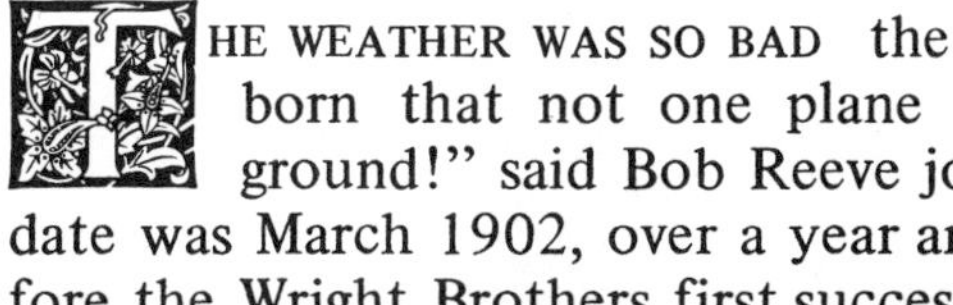

THE WEATHER WAS SO BAD the year I was born that not one plane got off the ground!" said Bob Reeve jokingly. The date was March 1902, over a year and a half before the Wright Brothers first successful heavier-than-air flight at Kitty Hawk. Robert C. Reeve and his twin brother, Richard, were born in Waunakee, Wisconsin. Two years later, their mother died and thereafter, their father remarried.

At an early age, Bob was restless and preferred hunting, trapping and dreaming, rather than schooling; formal education offered no stimulation. An avid reader with a photographic memory, Bob read everything he could obtain on his favorite subject: flying.

At age fifteen, he left home to enlist in the Army in World War I. When turned down by the first recruiter, he spent the last of his money traveling to the next town's enlistment station where he was accepted. Later, Richard enlisted in the Navy.

At the urging of his father, Bob returned to his home following his discharge with the rank of sergeant at the end of the war. His stay was brief and he ran off to San Francisco, where he signed on as a seaman aboard a ship sailing for the Orient. Jumping ship at Shanghai, Bob pursued his quest of self-education, drifting about the Orient for several years.

Returning home in 1921, he completed his high schooling in six months and entered the University of Wisconsin in the fall of 1922, where Richard was enrolled as a sophomore. During his two and one-half years of college, besides getting an education, Bob learned how to make gin, was a "fancy-Dan" with the girls and he dreamed of flying; the latter receiving priority.

After leaving college, he drifted to Beaumont, Texas, in March 1926, after a brief stay in Florida. In Beaumont, he joined up with barnstormers Hazard and Maverick, who traveled the country putting on an aerial circus. In exchange for support work, the pair gave Bob flight instruction and, after about three hours or so, told him he was ready to solo.

Opposite: Robert C. Reeve pauses for photo in his nostalgic Reeve Aleutian Airways' office in Anchorage, where he holds "court" with famous personalities, old friends and tourists, alike. He is, as his Stateside friends remark when they introduce him, "Mr. Alaska." Below: Present-day routes of Reeve Aleutian Airways.

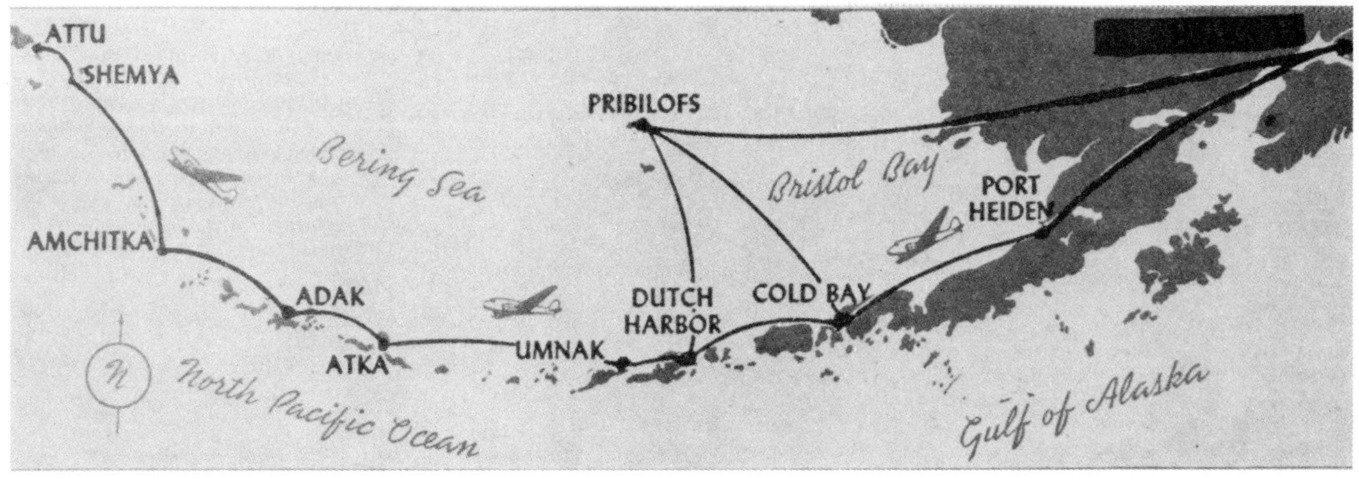

Above: Lamar Nelson, left, A. D. Smith, center, and Bob Reeve pause beside Kinner-powered Fleet biplane at Phoenix, Arizona airfield. Reeve completed his multiengine pilot training at Ford Motor Company tri-motor training school shortly after this 1928 photo was taken.

Right: Early in 1929, Reeve went to Peru in air-minded South America to assemble the $67,000 Ford Tri-motors and train PAA pilots. On an introductory flight to Peru's capital city of Lima, Reeve mistranslated an instruction telegram which read in part, "sin publicidad," and buzzed President Leguia's palace with the huge transport. An international incident was avoided when Leguia refused to recognize the low-level acrobatics.

Reeve Photos

Left: Reeve and his mechanic, Fred Rodriguez, pose for snapshot with fast, modern Lockheed Vega of Pan American Grace Airways, which Reeve flew after completion of the Ford project. Reeve, however, favored the new Fairchild 71 for the required high altitude flying.

Later, Bob did all of the repair work and tuning of the circus stunt plane. In 1926, when the government licensed pilots and mechanics, he received one of the first Engine and Aircraft Mechanic's licenses, as well as his commercial pilot's rating. Thus launched his career in commercial aviation, beginning with his South American flying in early 1929.

Much excellent documentation of Alaskan aviation history has been published, which includes the contributions made by many pioneer flyers, including Reeve, as of this publication date of 1971. Bob, while still actively engaged in commercial aviation in the 49th State, for years has unselfishly contributed to the archives and documentation of this important period of time. An accomplished writer, he has written extensively about his contemporaries, as well as being highly instrumental in contributions to the bush pilot Hall of Fame at the University of Alaska.

Reeve Photos

To Reeve, flying was not all aviation. Here he is being initiated into the Grupo Lanceros, a Chilean cavalry outfit stationed in Arica, Chile. Reeve says one of them ended up under the table, but it wasn't him!

Reeve is visited by his wife, Tillie, and son, Richard, after one of his famous mud flat landings at Valdez with his ski-equipped Fairchild 51. Reeve, in the mid-1930s, serviced Chugach Mountain Range gold mines with dangerous high altitude glacier landings that earned him the well-known identification "Glacier Pilot," which is the title of his biography by Beth Day. In modern days, the champion Anchorage baseball team is named "Glacier Pilots" after Reeve.

Reeve Photos

Prospector Reeve, in search of a gold strike, spent many off-hours flying to heretofore unaccessible promising areas. In photo at right, he conducts an analytical inspection of gold ore he collected from various locations throughout Alaska.

Reeve Photos

Upper: Members of Bradford Washburn's 1937 Mount Lucania expedition, Norman Bright, lower, and Russell Dow, upper, refuel the Reeve Airways' Fairchild 51 just prior to Reeve's record-breaking high altitude landing. The 8,650-foot Walsh Glacier landing was declared the highest landing for a ski-equipped aircraft at that time.

In lower photo, Reeve's "DGA" Fairchild 51 (damn good airplane), sunk its left ski in decomposed ice during an attempted take-off from Walsh Glacier. After a five-day wait for colder temperatures and stronger ice, Reeve, with full throttle, kangarooed down the rough, deep ridges of the packed ice, dropped off the edge out of sight and became airborne mere feet above the valley floor below.

World-renowned explorer and mountain climber Bradford Washburn, himself a pilot, is pictured above just before he and a companion, Robert Bates, accompanied Reeve on the flight to the base of then unscaled 17,150-foot Mt. Lucania of the Saint Elias Range. Washburn and Bates had to walk the 175 miles out when deteriorating glacier ice prevented Reeve from further landings.

Gordon Williams Photo

Boeing 80-A, NC 224M, dubbed "Yellow Peril," is shown above soon after Reeve ferried the Hornet-powered trimotor from California to Nabesna, Alaska, the summer of 1941. Reeve was under contract with Morrison-Knudsen Company to airlift supplies and personnel into Northway, an air base under construction by the CAA. By October 1941 Reeve had completed the task, often flying from dawn to dark, sometimes without engine shutdown.

Reeve Collection

Above: Reeve's only crash-landing occurred with his Boeing 80-A on a July 1943 flight from Anchorage to Amchitka. Late in the day, with radar men and equipment aboard, Reeve found his first scheduled stop at Cold Bay fogged in. After several unsuccessful attempts to land, he tried to set down in a nearby emergency field. Failing that, he pointed his plane toward the coast where experience had taught the veteran pilot that rifts in the weather could show above the shoreline. Finding an irregularity in the cloud blanket, he put the biplane into a gentle glide as one engine quit out of fuel. The ancient yellow Boeing hit the shallow surf "like a ton of bricks," upending the aircraft. All except the co-pilot were injured. It was after midnight when Reeve, notwithstanding an injured back, saw to the removal and care of his passengers on the nearby beach and sent out a position report from the salvaged radio. Rescued by a Coast Guard PBY the next day, Reeve remained at the site until his valuable cargo had been transferred to a crash boat. The huge Boeing had flown its last flight.

The 1947 photo below of same Boeing 80-A, sans engines, was long a familiar landmark at Anchorage's Merrill Field until mechanics sawed it up and hauled it to the dump. Incensed, Reeve ordered every last man to retrieve the antique and the plane is now being restored by the Pacific Northwest Aviation Historical Foundation at Seattle, Washington.

Kellogg Photo

N. Y. Daily News Photo

Reeve and son, Richard, with Boeing 80-A, left, and reliable red Fairchild 71, right, represent the "Air Force" of the Alaska Communication System (a branch of the Army Signal Corps), in 1942 photo above at Merrill Field. The only bush operator under military contract, Reeve flew men and supplies to ACS bases all over Alaska, western Canada and was the only civilian pilot assigned to a combat zone, the Aleutians; "birthplace of the winds."

Reeve Photo

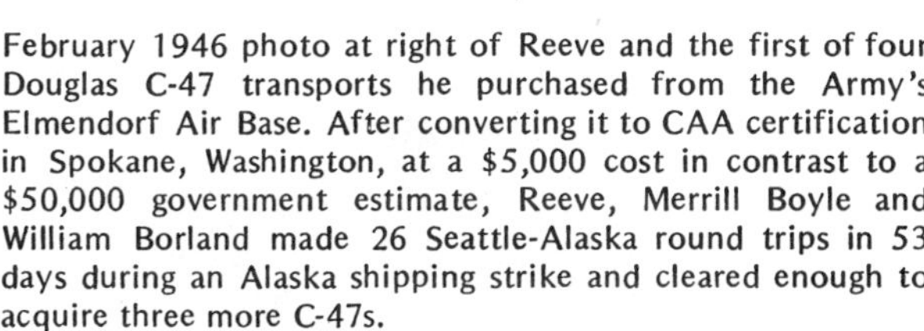

February 1946 photo at right of Reeve and the first of four Douglas C-47 transports he purchased from the Army's Elmendorf Air Base. After converting it to CAA certification in Spokane, Washington, at a $5,000 cost in contrast to a $50,000 government estimate, Reeve, Merrill Boyle and William Borland made 26 Seattle-Alaska round trips in 53 days during an Alaska shipping strike and cleared enough to acquire three more C-47s.

Reeve poses above with two of his three sons, Richard (standing) and David, before a painting of Boeing B-17s repelling Japanese forces during the early days of the Aleutian campaign. Below is wife, Tillie, and four of their five children: Roberta, David, Richard and Janice. The youngest son, Whitham, was born in 1947.

Opposite: A reception by the city of Anchorage of Northwest Airlines' 1946 survey flight for their newly certificated Orient route resulted in a reunion of old friends, NWA vice-president Frank C. Judd, left, Reeve, center, and NWA operations manager John F. Woodhead.

Reeve Photo

Photo by Steve McCutcheon

On a 1948 moose hunt near Farewell Lake, friends pause for photo at right. Left to right: Air Force Chief of Staff General Hoyt S. Vandenberg; wolf trapper Einar Carlson; famous flyer Colonel Brent Balchen; Reeve, and Commanding General Alaskan Air Command General Hamp Atkinson.

Reeve Photo

Below: World traveler and avid big game hunter Samuel L. Savidge (left of center), Seattle automobile dealer, is greeted by Reeve in Anchorage preparatory to flying to campsite of Kodiak bear hunting party.

Hayes Collection

Opposite: Reeve has just completed a check ride in a Convair F-102 in 1960. An honorary major in the United States Air Force, he was a flight commander of the Elmendorf-based 317th Fighter Interceptor Squadron until it was deactivated in 1969. The major was reassigned to the 43rd TFS in his old flight commander position.

In 1956 photo below, Reeve and his first Douglas DC-4. Even today, Reeve owns "every nut and bolt in the outfit." Strict safety and operational procedures invoked by him resulted in firing himself as a pilot. The airline's accelerated aircraft maintenance program is rated as having one of the lowest number of engine failures between engine changes of any carrier in the business.

Reeve Photo

Reeve Photo

B-24s return from a raid on Paramushiro. Based on a research project of Lt. Lawrence Reineke, intelligence officer of the 21st Bomb Sqd., permission was granted by the War Department to hit by air from the Aleutians. On July 18, 1943, six B-24s flown by crack 11th Air Force pilots, successfully raided the Japanese Paramushiro fortress. Flight leader Maj. Robert E. Speer, Maj. Edward C. Lass and Capt. Jacques Francine bombed the air base, while flight leader Maj. Lucian K. Wernick, Maj. Frederick R. Ramputi and Maj. Richard Lavin saturated harbor shipping. The 1700-mile mission shook Japan to revised defenses.

U. S. Air Force

SHEMYA ARMY AIR BASE

Important Rock In The Aleutians

CHAPTER 8

AT THE WESTERN END of the Aleutian chain lies the Near Islands Group, which includes Attu, Agattu and the Semichi Group, Nizki, Alaid and Shemya, which is the largest. About four square miles in size, Shemya rises abruptly from its northern shoreline to an elevation of 200 to 240 feet, and along the southern shore has an elevation of 40 to 80 feet.

Shemya was first considered militarily in the early spring of 1943, although the entire Aleutian Islands were almost ignored in pre-World War II American defense planning. The 1943 planning involved Shemya's role in relation to Kiska and Attu reoccupation.

As a result of an Eleventh Air Force order, weather aircraft were instructed to circle Shemya to ascertain whether it was Japanese occupied. With the belief that Shemya was deserted, an advance party of nine officers and eight enlisted scouts landed on the island May 28, 1943. Two days later, 2,500 troops under the command of Brigadier General John E. Copeland landed on Shemya. Immediate construction of runways got underway and planning of Shemya as a secret air base for the Boeing B-29 Superfortress was implemented.

Shemya's location west of the 180th meridian placed it closer to the center of east Asian population than any other United States territorial military station during World War II. It was important for various reasons: it acted as a base of operations in bombing raids against the Kurils; it was utilized as a transit base by Russian pilots ferrying planes and material under the Lend-Lease program; and, it provided essential Army and Navy weather stations and denied Japan a base for attacks against the United States.

After World War II, the Air Force deactivated Shemya to a small, custodian level and in 1955, Northwest Orient Airlines leased the island until 1958, when it was reactivated by the Air Force.

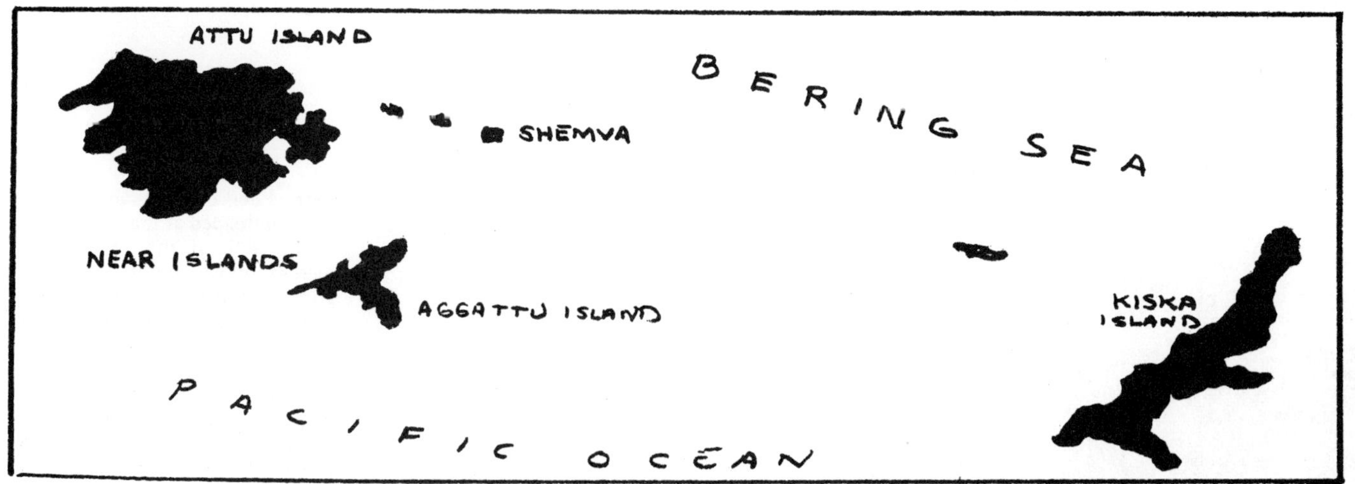

Left: SS "Scotia" rips her hull and sinks in treacherous Shemya Harbor, December 22, 1943, in a courageous attempt to deliver vital construction materials and equipment. Water transportation was the life line of Shemya. Prior to the dock facility completion, supplies had to be barged to shore.

U. S. Air Force Photos
(Except Where Noted)

Above and Right: Following the completion of the important preliminary airstrips on Shemya, a 300-foot crib dock and 1,400-foot ship dock was completed on July 25, 1944. Two breakwaters were later constructed; the longest stretched 3,500 feet.

Left: Fuel oil and aircraft fuel being unloaded at main dock. The rusted hulk of the SS "Scotia," seen in photo center, serves as a grim reminder of the mariner's dilemma. Attu, 40-miles distant, was set up a material staging area. The fantastic supply accomplishment was dimmed somewhat by a quartermaster's priority memo, "The toilet paper has not as yet arrived."

Right; Construction and earthmoving gets underway on first of three runways. The first phase was completed on June 21, 1943, less than one month after the original landing.

Below: Engineers and workmen lay down steel matting on main runway at Shemya. On June 24, 1943, the first plane, a Douglas C-47, landed at the new air base. The first heavy aircraft was a B-24 piloted by Captain I. L. Wadlington in August 1943.

Right: Hangar Number Four of the 11th Fighter Squadron is well under construction. The first flight of P-40s arrived at Shemya on July 23, 1943, and ten P-38 aircraft of the 54th Fighter Squadron, 343rd Fighter Group, arrived on August 10th from Amchitka.

Upper aerial photo of the island of Shemya shows main 10,000-foot runway that was built for planned Boeing B-29 operations against the Japanese homeland. Runways "B" and "C" each were 5,000 feet in length. All three strips were completed and in use by December 1944.

Left: Crew of water-cooled machine gun scan the horizon for Japanese raiders of Shemya. The last enemy air strike of the Aleutian campaign occurred on October 13, 1943, when nine four-engine bombers hit nearby Attu incurring little damage. Shemya defense forces were alerted for suicidal commando-type raids to destroy air base facilities.

An Army C-47 transport was the first plane to land on Shemya on June 24, 1943, less than 27 days after the U. S. occupation of the island. Eventually, a "YM" radar beacon aerial navigation aid was installed, in addition to a Bartow lighting and approach system.

Upper: North American B-25 medium bombers stationed at reoccupied Attu are sister ships of famous Doolittle Raid on Japan April 1942.

Right: Col. George A. Bassett, first commander of Shemya Army Air Base. Bassett was a member of the original scouting party that landed on the island and planned, developed and directed the construction of facilities at Shemya.

Above: D Model B-24 Liberator of the Shemya-based 404th Bombardment Sqd. Capt. I. L. Wadlington landed the first heavy aircraft on Shemya August 13, 1943, after circling several times waiting for heavy construction equipment to clear the runway.

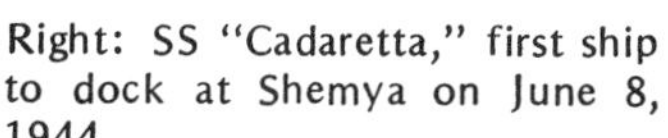

Right: SS "Cadaretta," first ship to dock at Shemya on June 8, 1944.

Left: Col. Robert H. Herman, Commanding Officer 11th Strategic Air Force, left, plans a mission with Lt. Col. Bryant, center, and Maj. Herold. First Shemya bombing mission was on March 16, 1944, off the island of Onekotan, located southwest of Paramushiro.

Below: Flight crewmen swing the tail of their B-24 in preparation for engine start-up.

Right: One of the B-24s of a six-plane flight taxis out for take-off.

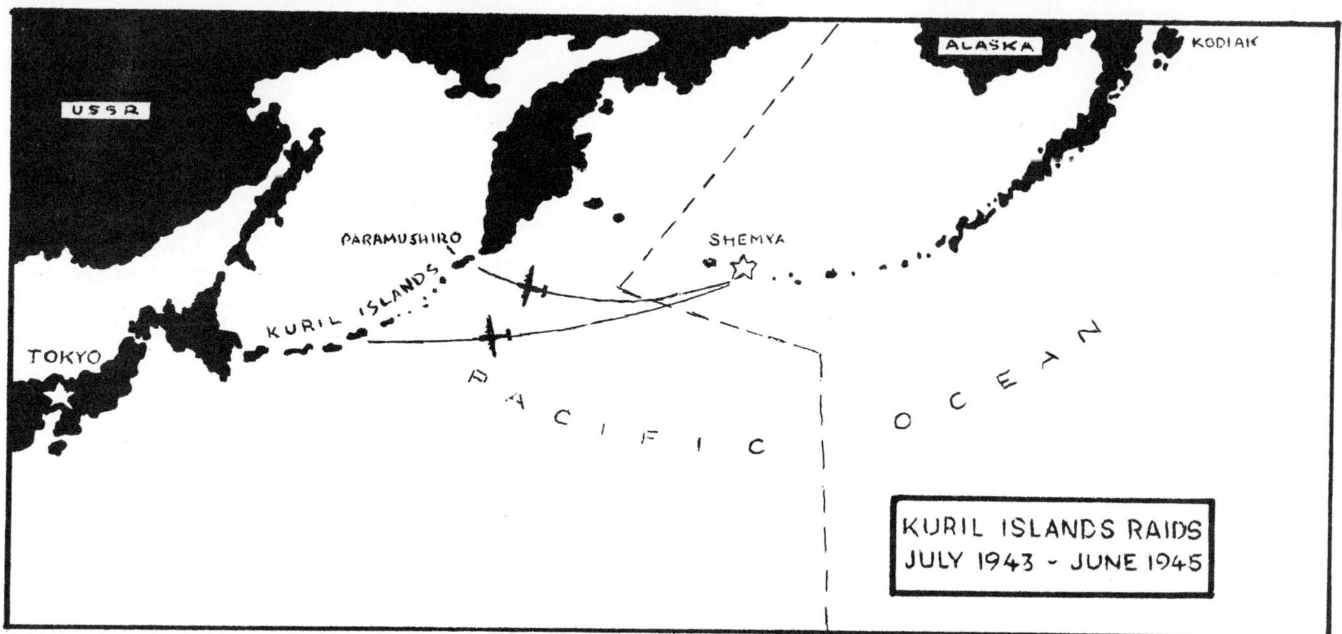

Left: Ground crewmen watch crippled flak-riddled B-24 of the 404th Bombardment Sqd., as pilot attempts to control landing following 11-hour long overwater mission.

Right: After debriefing the mission, B-24 crewmen receive their well-earned shot of "grog." Their day had been a long and strenuous one. Typically, they rise at 03:00, breakfast, then briefing by 05:00; to the flight line for engine start-up and take-off by 07:00, which begins eleven to fifteen hour engine-drumming flight, interrupted briefly by a "K" ration lunch. With keyed-up suspense over the target, swivel-headed gunners scan for enemy fighters; bombardiers set course, rate and drift to make their devastating cargo count; navigator hovers over charts computing figures, constantly monitoring their location; and pilots maintain formation in flak-turbulent air, flying the bombers straight and level on the bomb run. "Bombs away," and with enough fuel and favorable weather, they will make it home – mission complete.

Flak-damaged B-24, at left, limped home from Paramushiro bombing mission August 19, 1944, but crash-landed at Shemya, exploded and burned.

Crewmen point out fighter and flak-inflicted holes in B-24s of the 404th Bombardment Squadron. Flight line crews often worked around the clock in bitter weather readying aircraft for next mission.

Pilots of Shemya-based Curtiss P-40s and Lockheed P-38s of the 343rd Fighter Group distinguished themselves in meritorious combat engaging enemy fighters and bombers, strafing and bombing enemy vessels and land installations, and escorting bombers. In photo at right, a once operational "Warhawk" is reduced to rubble after a crash landing at Shemya. Below: More fortunate pilot saved his P-40, sustaining only a bent propeller at the precise edge of the base's 10,000-foot runway.

Left: Col. George R. Bienfang succeeded Col. Bassett as Air Corps Commander of Shemya Army Air Base on July 5, 1944.

Right: New Commander of the Eleventh Air Force Brig. Gen. Harry M. Johnson, right, is greeted by Col. Bienfang and Col. Goodman in December 1944.

Above: Visiting movie celebrities: Martha O'Driscoll and Errol Flynn, left, in Officer's Mess at Shemya, December 1944.

Above right: General Johnson briefs famous aviation figure Eddie Rickenbacker on the operations of Shemya.

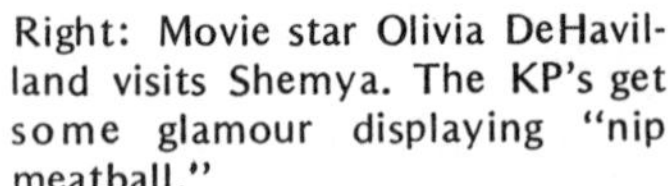

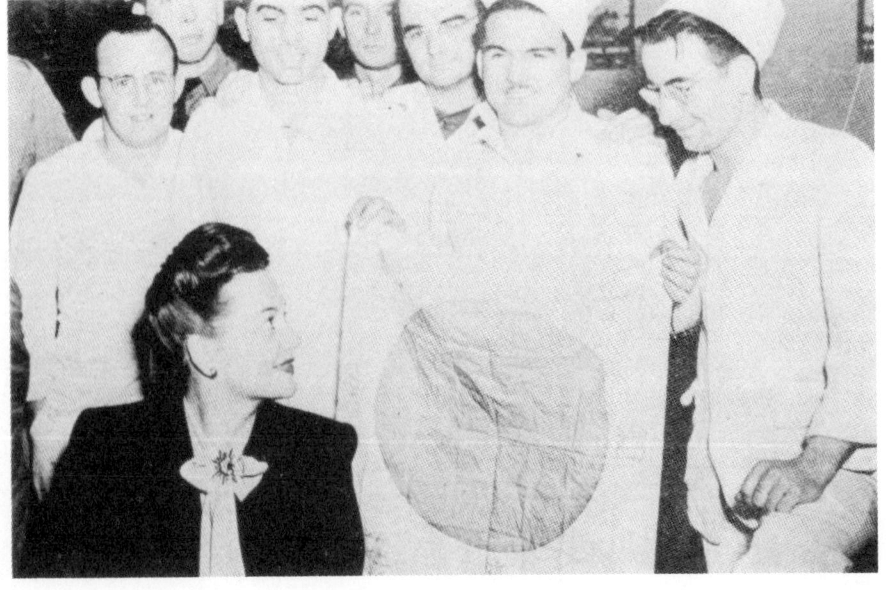

Right: Movie star Olivia DeHaviland visits Shemya. The KP's get some glamour displaying "nip meatball."

Left: Shemya harbor damage occurred during a raging storm November 13 and 14, 1944. The pounding surf leveled the entire east jetty of the breakwater, and 2,500 feet of the west jetty was carried away. The ships dock was completely gone and only 1,000 feet of the crib dock remained. The damage proved almost disastrous to an island wholly dependent upon shipping for its provisions.

Below: As a result of the devastating storm, responsibility of maintaining an open supply route fell to the Air Transport Section C-47s stationed at Shemya. Shuttle runs between Attu and Shemya were made by three aircraft and an average of 21 flights per day were made.

Above: Troop replacements arrive at Shemya Air Transportation dock after the harbor deactivation by storm. C-47s flew supplies and troops, as well as saving Shemya's effectiveness. Left, upper and lower: C-47 crash-lands at the end of the runway.

Right: On August 20, 1944, paving of the runways at Shemya began with specifications to handle wheel loading of Boeing B-29s. The contract was awarded to S. Burch & Company and Morrison-Knudsen Company, Inc., in order to complete the project as rapidly as possible. The world's largest portable asphalt plant was built on the site, shown above.

Below: The last bombing mission flown by the 11th Air Force was from Shemya to Paramushiro on August 13, 1945. From January to August, 1945, a total of 393 tons of bombs were dropped by the 404th. Of this tonnage, 56.3% were by planes utilizing the airborne radar system. The primary targets were Katacka Naval Base on the Island of Shimishiru. In addition, in the Kurils, Paramushiro and Shinush were attacked.

Upper: The first B-29 aircraft landed at Shemya on May 11, 1945. This aircraft was assigned to the cold-weather station at Ladd Field and came down the Aleutian chain on a "Morale Flight." The giant superfortress landed at 10:27, five minutes before it was scheduled. Other B-29s were not seen until September 16, 1945, when four of them landed after a non-stop flight from Iwo Jima in the Central Pacific.

Below: Lockheed Constellation of Northwest Orient Airlines. In 1955, Shemya was leased to Northwest and the company operated the field until 1958, when the Air Force reclaimed the island and established an Air Force Station. Its mission continued to expand with the Cold War and in 1968, due to the Vietnam Conflict, it became again an Air Force Base. Today, it supports Army, Navy and Air Force activities, mostly classified secret.

Northwest Orient Airlines

Morrison-Knudsen Archives

CAP: ALASKA'S WAR BABY

Futile Air Search For Harold Gillam

CHAPTER 9

FROM THE BEGINNING of his early flying over Alaska, former heavy equipment operator Harold Gillam demonstrated an uncanny ability to master the adversities confronting all flyers earning a living in the Northern wilderness.

Since his first solo flight in 1928, some said he was crazy; others that he was lucky. Time demonstrated that Gillam, with luck but more important an inherent skill, developed a widespread reputation for getting through. After fifteen years of conquering the air space over Alaska, the veteran pilot's luck ran out. His skill even then was never in question.

On January 5, 1943, Gillam, piloting a Morrison-Knudsen Lockheed twin-engine Electra with five passengers, was missing on a flight from Seattle to Ketchikan. Every possible means of search was being utilized by Morrison-Knudsen, Coast Guard, Army, Alaska Game Commission, civilian pilots and the Canadian Air Force.

An hour out of Ketchikan, in a violent storm, one engine quit and the Electra started to lose altitude. For life-long minutes, peaks and trees skimmed by and the plane shuddered as one wing ripped off trees which slammed the small airliner into a snowclad mountainside. The only woman aboard, Susan Batzer, was gravely injured and died two days later. Dewey Metzdorf of Anchorage, Percy Cutting of Hayward, California, Robert Gebo of Seattle, and Joseph Tibbets, CAA communications supervisor, all received various degrees of injuries. Gillam, with internal injuries, left two days later down the mountain for help.

After a futile air search in which planes flew so close to the crash site that the survivors could make out the pilot's features, Tibbets and Cutting followed Gillam for help. The tragic 30-day ordeal closed when Tibbets and Cutting were rescued, followed by Gebo and Metzdorf. Gillam was found dead at the bottom of the mountain – one mile from the rescue contact point. One of the pioneer all-time great flyers had flown his last flight.

Opposite: Charles "Harold" Gillam slips into cockpit of his Swallow biplane at Fairbanks in 1929, shortly after he learned to fly. Early in his flight training, Gillam escaped death when he and his instructor, H. L. Danford, crashed near Fairbanks. Danford was the victim of the first fatal air crash in Alaska. Below: Morrison-Knudsen Co. Lockheed Electra which, piloted by Gillam, crash-landed near Ketchikan, Alaska, January 5, 1943, with six people aboard. A woman passenger died as a result of the crash and Gillam died later from exposure after a vain effort to seek help.

Morrison-Knudsen Archives

In 1929, Gillam flew this Stearman plane on Eielson search. A novice pilot at the time, Gillam was with veteran flyer Joe Crosson when they found Eielson's crashed Hamilton. Later the fearless flyer operated Gillam Airways out of Copper Center where he acquired the infamous title, "Bust-em, fix-em-fly-em Gillam."

Lewis Collection

Below: Ireland "Neptune" amphibian of Gillam Airways flipped over on its back at Merrill Field, Anchorage, in 1931, when one landing gear failed to lock down.

Mills Collection

Gillam's trim, little Zenith biplane, at right in 1931, operated out of Copper Center until 1935, when he moved his operations back to Fairbanks. Later, Gillam refused to have airway signs painted on his planes.

Ed Young Collection

Photo below is of shy, moody Gillam, right, and friend at Boeing Field, Seattle, where the well-liked bush pilot took instrument instruction in the late-1930s.

Alice Harris Photo

Gordon Williams Photo

American Pilgrim 100-A, at left, during stopover at Boeing Field in July 1935 on flight to Fairbanks. The single-engine plane was put into service briefly with Pacific Alaska Airways. Below is same plane seven months later on iceberg-infested tide flats near Anchorage. Gillam, who had recently purchased the sturdy aircraft, circled fog-shrouded Anchorage looking for a clearing to come down through when he ran out of fuel and landed among the ice floes of Cook Inlet. Uninjured, the passengers and pilot climbed onto the wing and were rescued almost immediately.

Mills Collection

Below: Gillam prepares for his mail run from Sleitmut on the Kuskokwim River. On one such flight, the nerveless aviator landed at his scheduled postal stop in a blizzard, delivered the mail pouches, refueled his plane and took off to the consternation of three veteran bush pilots who had been grounded for days waiting for clearer weather.

Alice Harris Photo

Right: U. S. Coast Guard pilot searches for missing Gillam Electra with six people aboard in January 1943. With only meager clues, the futile fifteen day search covered most of the rugged peaks of southeastern Alaska and western British Columbia.

Below: Army B-25 bomber joins U. S. Coast Guard and Royal Canadian Air Force in all-out search. Several times rescue planes flew so close to the crash-site that the survivors could see the pilot's heads turning.

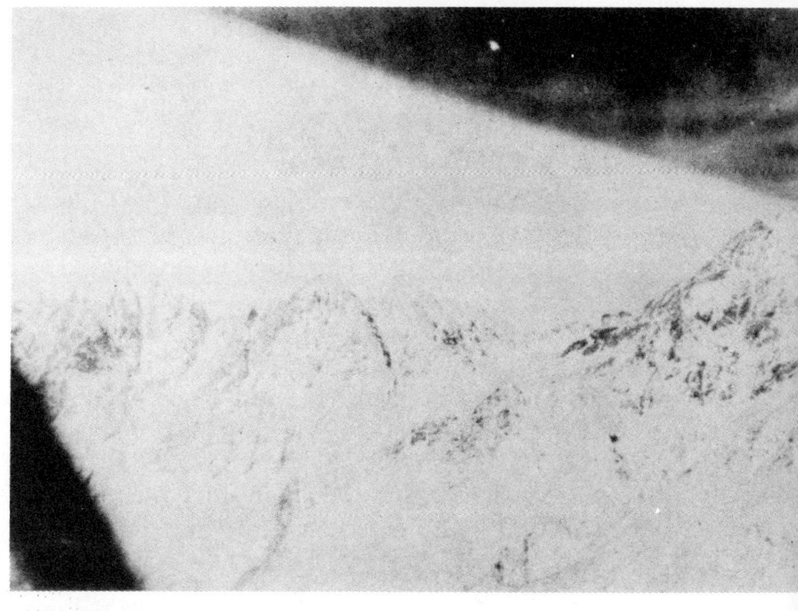

Waterworth Photo

U. S. Air Force

With Alaska's pre-war establishment of flying as the main mode of transportation and the subsequent military activity during and following World War II, it was evident that air search and rescue was an important necessity in air-minded Alaska. Besides the Air Force's highly trained rescue squadrons based throughout the vast North country, the activation of the Civil Air Patrol in Alaska was of prime importance.

The Civil Air Patrol (CAP) is the child of early World War II days in which many civilian pilots, otherwise unacceptable for war duty, offered their services flying supplementary missions in support of the war effort. It may be said that the advocates of a nationwide CAP made numerous unofficial overtures to U. S. government authorities in an effort to establish their organization as an element in the National Defense.

Civil Air Patrol

Spearheaded by New York Mayor Fiorello H. LaGuardia, Director of Civilian Defense, an order was signed on December 1, 1941, creating the CAP just one week before Pearl Harbor. With such wartime duties as coastal patrol and courier service, the CAP meritoriously served America.

The Civil Air Patrol in Alaska was established on April 21, 1948, with headquarters at Anchorage. Some of the early charter members were Mason LaZelle, John Manders, James Carter, Bob Reeve, Virgil D. Stone, George Weyer and Jack Carr, who became the first wing commander.

The Alaska Wing CAP was organized under General Order No. 3 dated 21 April 1948 as an auxiliary of the United States Air Force. With one aircraft assigned and many member-owned planes, CAP in Alaska soon was recognized as a vital reassurance to the swelling civilian pilot population. Strictly on a dues-paying volunteer basis, members of the wing additionally trained in ground rescue including recovery, first-aid and evacuation.

An equally important mission of the national organization, the cadet training program was adopted in the early 1950s.

Elmendorf AFB Pho

Left: Photo of Beechcraft AT-17 of Fairbanks-based 10th Air Rescue Squadron.

Carter

Above: Colonel James E. Carter, Civil Air Patrol, Alaska Wing Commander. With headquarters in Anchorage, Carter has served as commander since 1959 and is one of the charter members of the Alaska Wing. In 1970, he received the Wing Commander of the Year award in Washington, D.C.

Above: The emblem and shoulder patch of the Alaska Wing, CAP. Below: Original aircraft assigned to CAP at Anchorage was Stinson L-5 "Sentinel" and Aeronca L-16. Lacking resources to refinish their planes in CAP markings, the surplus aircraft were flown with Air Force numbers and insignias, as shown below.

Alaska Wing CAP

U. S. Air Force

Above: Civil Air Patrol National Commander Brig. Gen. Richard N. Ellis, USAF, assumed his post on October 3, 1969, with headquarters at Maxwell AFB, Alabama. A native of Salt Lake City, Utah, Ellis has served over thirty years as an Air Force career officer.

CAP

U. S. Air Force

Above: Colonel William R. Stickman, Jr., U. S. Air Force liaison officer for the Pacific Region CAP, with headquarters at Hamilton AFB, California. The largest of eight regions, the Pacific Region includes Alaska, California, Hawaii, Nevada, Oregon and Washington. Stickman, a career officer, has held his current post since 1969.

Photo below depicts one of the important missions of CAP, the cadet program. With the availability of scholarships and national flight training programs, youth between the ages of 13 and 19 are trained and educated locally in aerospace and flying.

Lt. Col. Virgil Stone, deputy Alaska wing commander, right, briefs pilots Capt. Bob Mason, left, and Bob Claypool (hidden) on area of 1954 search. Alaska, the largest state, with more airplanes per capita in the United States, makes Alaska Wing one of the most active search and rescue organizations nationally.

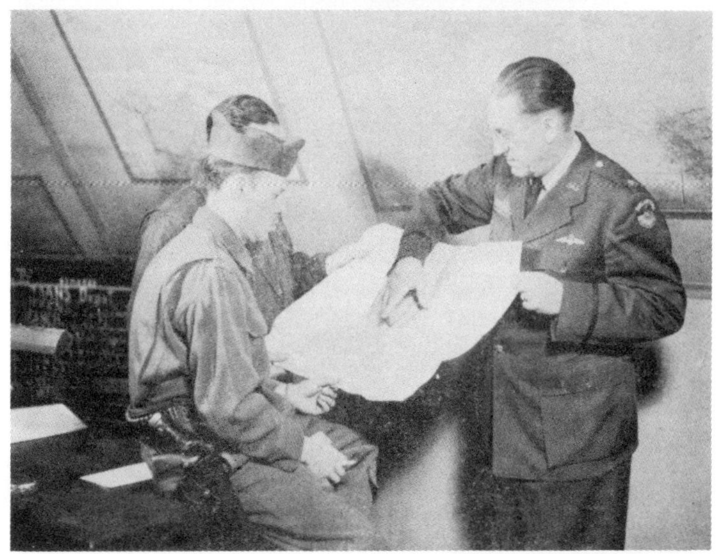

Alaska Wing CAP Photos

Above 1957 photo is CAP Stinson L-5 being defrosted and engine-warmed prior to flying a search mission. The L-5 served well as a search aircraft in the early days of the Alaska Wing. The rugged terrain required the horsepower and maneuverability of the World War II liaison plane.

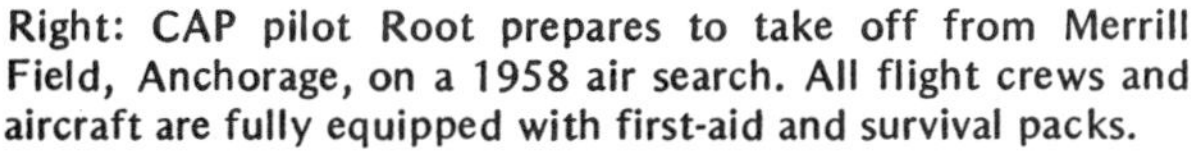

Right: CAP pilot Root prepares to take off from Merrill Field, Anchorage, on a 1958 air search. All flight crews and aircraft are fully equipped with first-aid and survival packs.

Alaska Wing CAP

Left: Colonel Harry E. Stiver, Alaska wing commander in 1958, climbs aboard a CAP Navion L-17. Stiver joined CAP in 1949.

Below: One of Alaska Wing's old favorites, "49 Tango" Super Cub. This plane is still being built today, attesting to its wide acceptance as a dependable and extremely airworthy aircraft.

Alaska Wing CAP

Below: Darwin Kellogg and Frank Munsey, flying the Cessna 180 shown at right, spotted Marc Stella's downed aircraft at 9:30 a.m. Wednesday, January 3, 1962. One of Stella's passengers, Warren Thompson, suffering from a rupture, was immediately flown to the 5060th USAF Hospital at Fort Wainwright, Fairbanks. Kellogg then flew back to the scene, accompanied by a mechanic and spare parts. Stella became the object of an extensive Civil Air Patrol search on Tuesday, after he failed to arrive at Fairbanks on a flight from Tanana. More than 15 aircraft were involved with the search.

U. S. Air Force

Right: On October 26, 1959, Major General C. F. Necrason administered the oath of enlistment into the Air Force to Civil Air Patrol Cadet Capt. William S. Ponge. Alaska's first CAP cadet to enter the Air Force, Ponge received the rank of airman third class.

U. S. Air Force Phot

Alaska Wing CAP

Left: Alaska's Governor William A. Egan, right, himself a pilot, congratulates Capt. Glenn Kipp, Kenai Squadron Commander CAP, on saving a woman's life on a mercy flight from Kenai to Elmendorf's 5040th Air Force Hospital. The event took place one week after Kipp was assigned one of ten L-20 DeHavilland Beavers released to CAP by the Air Force in March 1960.

Below: Alaska Wing CAP search pilot Gene Weiler, left, and the author's wife, Patricia Mills, a CAP member of Washington Wing, pause beside Piper Twin-Apache prior to orientation flight from Merrill Field over Anchorage and the Kenai Peninsula in March 1971.

Mills Photo

Ted Tax Photo

Left: DeHavilland L-20 Beaver, which is the basic aircraft used today in Civil Air Patrol Alaska Wing operations.

Alaska Airlines

POST-WORLD WAR II

Flying Becomes Alaskan Way of Life

CHAPTER 10

WITH THE REMOVAL OF World War II purchasing and travel restrictions Alaska, as elsewhere, quickly returned to a peacetime status. The airplane, for over two decades a recognized vital mode of transportation, assumed popular acceptance by the majority of Alaskans as a necessary extension of their environment as private pilot aircraft owners.

Five years of wartime military flying in the North country produced residules for private aviation in an already air-minded territory such as airports, navigational aids, radio communications and up-dated charts. As newer modern light aircraft became available, they appeared on the Alaska scene virtually overnight.

The vast, inviting territory was compressed as adventurers from all walks of life took to the air, reaping the grandeur and solitude of America's untapped wilderness. Sportsmen, mountaineers and explorers, along with naturalists, photographers, ardent rock-hounds and recluses escaped the humdrum daily toiling seeking self--fulfillment. With their aircraft, they "trucked-in" essentials to construct cabin retreats at placid locations hundreds of miles remote from the everyday world of city living. Descriptive new names of lakes, islands and estuaries became familiar as flyers referred to their favorite wilderness haunts.

Opposite: Air travel in Alaska, established as the main mode of transportation for almost two decades, witnessed a general aviation explosion immediately after World War II. Rugged, powerful Norseman is pictured loading for scheduled flight into central Alaska.

Modern-day bush pilots fuel one of three Alaska Airlines AT-19 Stinsons at settlement of Holikachuk, located in western central Alaska. Ever-present local Indian youth look on. Alaska Airlines, during this 1946-1947 period, was the largest bush flying operation in the Northern Territory.

Alaska Airlines

Since World War II hardly a section of Alaska had not been visited by the airplane, such as beach landing at left for fishing in remote Kenai area river.

Kellogg Photo

As a consequence, additional aircraft were acquired and pressed into service by such agencies as forestry, fish and wildlife, search and rescue and law enforcement to monitor their responsibilities to Territorial resources and airborne dispersed population.

The mushrooming aviation business, commercial as well as private, created a marked increase in Alaska's economy. New employment and finances resulted from flying schools; aircraft sales, parts and service; accessory sales such as skis, pontoons, winter-operation essentials and shelters, to name a few. Fuel and oil sales, repairs, airport fees, licenses and taxes, additionally contributed a flow of money brought about by growing airlines and commercial aviation, as well as general aviation's private flying.

With its geographical location, Alaska soon became the gateway of the air age after V-J Day and Anchorage, the air crossroads of the world as a refueling and connection stop for international air carriers flying the polar and great circle routes. In 1960, soon after its dedication, the new International Airport at Anchorage became the fifth busiest air terminal in the United States, serving such airlines as Air France, Alaska, Japan, KLM Dutch, Northwest Orient, Pacific Northern, Scandinavian and other major airlines.

In 1970, Anchorage International handled over 2,387,600 total passengers and 69,000 short tons of cargo. That same year, the new huge modern terminal at the busy airport was dedicated in preparation for increasing passenger limits that forecasters predict will reach over 8,500,000 in 1990. Alaska – the air capital of the world.

Winchell Photo

Right: Many early time bush pilots earned a good living supplying Alaska's growing interior. Everything was traveling by air. Old pre-World War II bush planes like Oscar Winchell's Stinson, shown here, were pressed into service.

Wien Consolidated

Photo above shows loading operations of Wien Airways' Norseman for supply flight to U. S. Navy oil exploration crews on the North Slope of Alaska interior. With expanding air routes and charter flights, the pioneer air carrier acquired smaller air operators and was renamed Wien Airlines.

Sullivan

Right: Present Mayor of Anchorage (1971), Valdez-raised George M. Sullivan served in the Aleutians during World War II. Sullivan was stationed at Adak and during his tour received the first Army Commendation Medal issued in Alaska from Col. Johnson. The chief official of Alaska's largest city was discharged as a sergeant in 1946.

Left: Post-war bush pilot, Dave Kellogg, sits on trailing edge of his Travelair wing, dejectedly contemplating his dilemma. The plane's engine suddenly packed up over the Kenai Peninsula and the World War II bomber pilot made an emergency landing in a bog, flipping the 1929 monoplane over on its back. Sustaining only minor damage, the aircraft was retrieved later. Kellogg, a pilot's pilot, flew the bush for five years; at one time for three companies at the same time.

Kellogg Photos

One of Kellogg's extra moonlighting jobs entailed flying temporarily repaired wrecked airplanes out of the bush for Jack Carr, operator of an Anchorage aircraft repair facility. At right is Supercub, with a collapsed landing gear as a result of a rough snow landing. Below: This brand new Piper Pacer rolled over when the engine quit on take-off. Owner J. Vic Brown, Jr. was not seriously injured. Carr, who was engaged to repair the float-plane, bolted the engine back in, replaced the propeller, straightened the struts and patched the pontoons with tar and tin from empty gas cans. Kellogg flew the botchery back to Anchorage with a slight droop in one wing and a cable laced through the floor served as a throttle.

Right: Woodley Airways served the Kenai area with twin-engine Cessna in early 1950s. Dave Kellogg mastered beach landing by staying on the rough but high side. **Below:** Pilot Don Dorothy flipped his SR-9 Stinson as a result of its wheels sinking into the soft tideflats.

Kellogg Photos

U. S. Air Force

Right: The ever-ready 10th Rescue Squadron Beechcraft AT-7 answered many bush pilot's calls for assistance.

Left: Haakon "Chris" Christensen checks the Ranger engines of his Grumman Widgeon. "The Flying Dane" came to Alaska in 1933 and the following year set up operations at Cantwell, flying a Curtiss Robin. In 1940, he moved his business to Anchorage, where he developed a route between there and Seward. He sold out to Merle K. Smith of Cordova Airlines in 1952. Christensen was killed in a crash of his Widgeon while flying in a severe storm near Cordova.

Photos Kellogg Collection

Anchorage-based Norduyn Norseman of Christensen Air Service is being loaded for charter flight to Bristol Bay area. Below: Fire destroyed Alaska Airlines' huge hangar at Merrill Field the winter of 1949-50. The original hangar was built in 1933.

Thompson Photo

Above: Bette Thompson, with daughters, Gale, right, and Chris, enjoy a leisurely afternoon on Alexander Creek, where they had flown to in Francis Mayer's Cessna 170 floatplane.

Left: In early 1950s, Robert Jensen commuted between his radar site employment on Fire Island to nearby Anchorage in Jack Carr Flying Service's Cessna 195.

Bob Jensen Photo

Thompson Photo

Right: Ken Thompson, left, and Keith Young remove spruce bows from air scoops of their Stinson Stationwagon. They explained the foliage was picked up flying a little low while chasing a bear.

Alaska Airlines

Left: Early Alaska Bell helicopter, while flying in unstable air, lost its tail rotor section as a result of a crash in 1950. First introduced en masse commercially by Alaska Airlines in early 1950s, the unsuccessful venture belied present-day utilization of copters in air transportation and industry.

Below: The Alaskan lure, America's last natural grandeur as viewed from a typical cross-country flight. By 1970, Alaska had over 4,500 private aircraft and exceeded 8,000 licensed pilots. Lake Hood, in Anchorage, had become the world's largest floatplane base with more than 400 take-offs and landings per day.

Kellogg Collection

Above: Lockheed Electra made an emergency shallow water landing between Fire Island and the mainland near Anchorage in early 1950s. On a flight test for single-engine operation, electrical failure necessitated injury-free forced landing. Many small carriers, like Cordova Air Lines, answered short route travel demand with safe, dependable flight service. Veteran bush pilot Merle "Mudhole" Smith built the almost defunct air service into a creditable, solvent venture. Cordova Airlines sold out to Alaska Airlines.

Boeing Co.

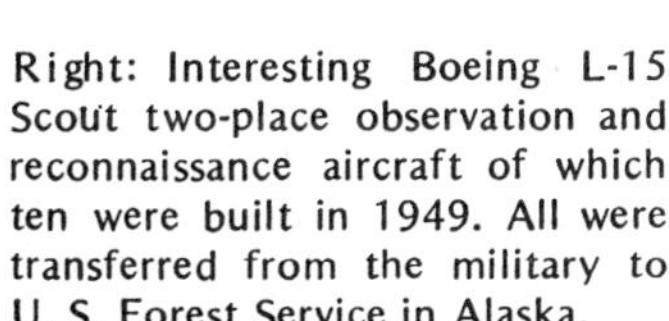
Right: Interesting Boeing L-15 Scout two-place observation and reconnaissance aircraft of which ten were built in 1949. All were transferred from the military to U. S. Forest Service in Alaska.

Reeve Collection

Above photo of renowned glacier pilot Bob Reeve's Fairchild 51 was taken in 1934 following a hazardous landing on Columbia Glacier, near Valdez, while servicing his "Ruff and Tuff" mine. In 1933, Reeve made his first Alaska glacier landing on nearby Brevier Glacier at the 6,000-foot level; in 1937, he made a record breaking 8,650-foot landing on Walsh Glacier of the Saint Elias Range. Reeve's ingenious glacier landing techniques later played an important role in similar high altitude operations.

Monsen Collection

On April 25, 1932, Joe Crosson, flying a Fairchild 71 similar to one shown at left, became the first pilot to land on Mount McKinley. Jerry Jones followed Crosson's Muldrow Glacier landing while the pair were supplying an expedition.

Sheldon Collection

World-famous bush pilot Don Sheldon flies off the 6,000-foot high icy slopes of Mount McKinley to return to his base of operations at Talkeetna, Alaska, 60 miles to the south. In 1960, Sheldon saved the lives of two mountain climbers when he plucked them from the 14,500-foot reaches of America's highest peak in two precarious trips. It established a record for the highest fixed-wing aircraft landing and take-off made on the mountain. Sheldon made his first McKinley landing in 1951.

THE MASTER OF MOUNT McKINLEY

Donald Edward Sheldon has gained the reputation of owning Alaska's Mount McKinley, but the ruddy bush pilot modestly claims only to have earned a piece of it. His investment began with his first glacier landing on North America's highest peak in the 1950s.

Today, 23 years of experience flying polar bear hunts in the Arctic, search and rescue missions in the rugged terrain and flying thousands of square miles of the bush qualifies him to glance up from the airstrip of his Talkeetna Air Service to Mount McKinley 60 miles distant and give a rundown on its weather conditions.

After using up 44 airplanes, of which three were wiped out, the cautious, finger-tip flyer has established his famous aviation business. His customers vary from delivering groceries to residents of the bush to glacier landing famous international mountain climbers on Mount McKinley of the Alaska Range and the Wrangell Mts.

Flying the most popular Sel-Aircraft equipped with retractable skis, floats and sandbar landing gear in 1965, Sheldon committed himself to $100,000 plus interest to purchase three Cessna 180 aircraft, and a new Supercub. With the new planes he retired the loan in five years, averaging over 1,000 hours flying per year.

Almost a legendary figure in his own time, the Master of McKinley is an astute businessman, devoted husband and father and doing what he likes best – flying.

Sheldon Collection

U. S. Air Force

Colorado-born Donald Edward Sheldon first came to Alaska in 1938 at the age of 17 and briefly drove a milk truck in Anchorage before moving into the bush north of there to Talkeetna. He worked at gold mining, skinned beavers and chopped wood to make a living. His exposure to the bush pilot's world while attending the University of Alaska, and subsequent travel around Alaska with a survey team, resulted in his taking flying lessons at Anchorage in late 1941.

In October 1949, Maj. Gen. C. F. Necrason, commander of Alaskan Air Command, presented Don Sheldon, a World War II veteran, the Air Force's highest peacetime civilian commendation, the Exceptional Service Award. Sheldon was cited for identifying, in December 1958, an ill-fated C-54 which crashed into Mount Illiamna, an awesome 10,116-foot peak in southwestern Alaska. Flying low through treacherous air currents, the lithe and wiry bush pilot made six passes over the downed aircraft before he was able to ascertain there were no survivors and make positive identification of the aircraft. Some of Sheldon's many other spectacular flying achievements included a July 1955 rescue of eight soldiers from a remote river for which he received the Army Meritorious Achievement award. In February 1954, he was involved in the rescue of three survivors of a C-47 crash near Mount McKinley. During World War II the celebrated pilot, as an Eighth Air Force "Flying Fortress" gunner, survived two crashes of B-17s shot down by Germans. His wartime commendations include The Distinguished Flying Cross and Air Medal with five clusters. Unimpressed by his fame, the modest glacier pilot emphasizes his good luck during over 23 years of flying the bush and majestic Mount McKinley.

Sheldon Collection

Located in the Alaska Range, Mount McKinley's 20,320-foot north peak was first scaled in 1913 by Archdeacon Hudson Stark's party. The 19,370-foot south peak was first conquered in 1910 by Pete Anderson and Billy Taylor. In the 1950s, Sheldon flew for explorer and Director of Boston Museum of Science Bradford Washburn, whose seven expeditions on McKinley and 15 years of research culminated in the intricate mapping of Mount McKinley in 1964. This opened up "Mighty Mac" to many expeditions. Since 1951, Sheldon has supported 100% of all first climbs.

On returning to Talkeetna in 1948, Sheldon bounty-hunted wolves at $50 each to make ends meet. At right, wing-mounted rifles were ingeniously fired from cockpit of his Supercub. The veteran pilot now limits his hunting to an occasional moose to meet his family's needs. The years of flying have transformed him into an ardent conservationist. He reflects about the vast bush country that once teamed with wildlife.

The first winter scaling of the north peak of McKinley occurred February 28, 1967, when Ray Genet of Anchorage, Art Davidson, Fort Morgan, Colorado, and Dave Johnson, Gladstone, N.J., endured winds of 150 mph and temperatures of -71 degrees to succeed. In summer, Mount McKinley is inhospitable; in winter, it is deadly. The ordeal began January 10, when the party of eight climbers headed by Gregg Blomberg were flown by Sheldon to Kahiltna Glacier, 7,250-feet up McKinley's west slope. Less than 24 hours later, team member Jacques Batkin from France, fell to his death in an ice and snow-hidden crevasse. On February 26, three of the seven climbers reached the 17,300-foot level. The next day, they tried for the summit but were forced back by a whiteout. The victorious three struck out and succeeded the following day. As they started back down, a raging storm was encountered at the 18,200-foot level. For six days, they huddled in a snow trench. The remaining four companions below gave the three up for lost, and worked back down to 10,000 feet. Stamping out a message in the snow, "WEX6-DON HELP," the party waited for air evacuation. A dramatic effort involving Army helicopters, Sheldon, and volunteer climbers resulted in the rescue of the three from the 14,000-foot level and four at the lower level.

All Photos Sheldon Collection

Don and Roberta Sheldon were married in May 1964, and have two young daughters and a newly-born son. Roberta is the oldest daughter of pioneer bush pilot Bob Reeve. While her husband functions as pilot, mechanic and even grocery clerk filling orders for bush customers, Roberta is the bookkeeper, secretary, radio operator and trouble-shooter.

Don Sheldon, center, Ray Genet of Anchorage, left, and members of Mount McKinley expedition led by Walt Gonnoson pause briefly for this mid-1950 photo.

Photos Sheldon Collection

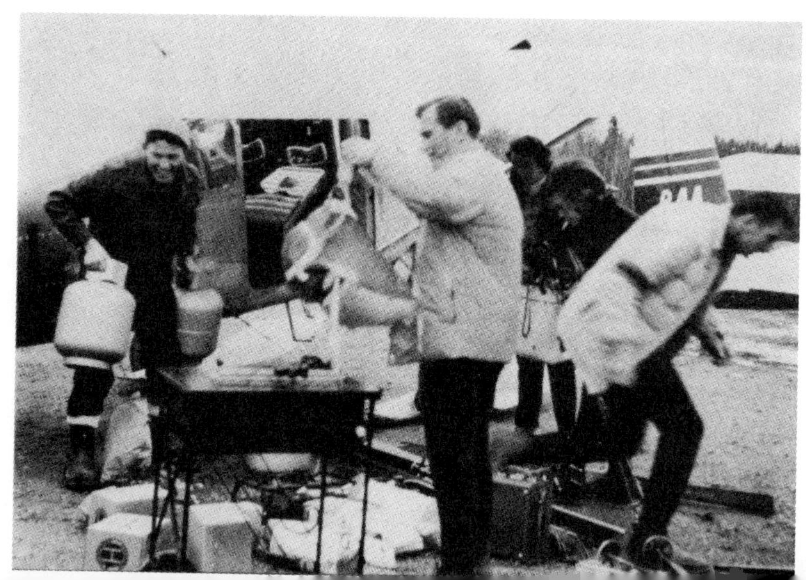

Sheldon, left, and party load one of his Cessna 180s for trip. Sheldon saves 20 pounds of weight by not painting some of his aircraft.

Sheldon's escapades have been many and varied; including North Coast polar bear hunts, wild and woolly adventures with cold weather testing, Army, Air Force and Institute of Biology research teams, and all flying and "crack-up" scenes for Warner Brother's movie, "Ice Palace." The master of McKinley regards his flying job as a perpetual vacation, but has had his share of misfortune. In 23 years of bush flying, he has owned 44 planes, three of which were "totaled" in accidents. His most serious injuries, however, occurred when he fell 400 feet down a mountainside when the shale he was traversing gave way. He saved himself from certain death by grabbing a rock protrusion. After rescue by his companions, Sheldon survived the 16 miles back to his plane, suffering from a broken shoulder and multiple body cuts and bruises.

A widespread search was launched for Sheldon when he was forced down, low on fuel, after battling severe storms on a flight from Pt. Hope to his intended destination of McGrath. After a skillful damage-free landing high on a mountain side, Sheldon transmitted his position on H.F. radio frequencies, keeping the aircraft engine idling for power and getting the last drop of fuel by lifting the tail of his plane. McGrath advised him to listen for sonic boom of Colonel Rodgers and his jet drivers and record the time and direction. A.D.F. and F.C.C. fixes pin-pointed the downed Cessna 180. A heavy multiengine plane air-dropped fuel after five days of bad weather searching, and Sheldon continued his trip.

Sheldon Collection

Northern Consolidated

Above: President John F. Kennedy, followed by Alaska's Governor Egan, boards Fairchild F-27 turbo-prop airliner on 1960 visit to Alaska.

Right: November 1958 photo taken at Anchorage International Airport of Vice President Richard M. Nixon; Mrs. Nixon, The Honorable J. L. McCarrey, Jr., Judge of the District Court, Anchorage, and Ray Petersen, president of Northern Consolidated Airlines.

Giant Douglas C-124 Globemaster outsize cargo is unloaded at Taltalina, Alaska, during Operation Shoehorn in May 1958. The huge transport of the 62nd Troop Carrier Wing based at McChord AFB, Washington, is on support assignment to the Alaskan Air Command. The mission required dangerous operations in and out of short, narrow and rough airstrips. The wing was redesignated the 62nd Military Airlift Wing in January 1966. They transisted to C-141 Starlifters in August of that year.

U. S. Air Force Photos

At first glance, this C-124 Globemaster appears to be the leftovers of a mountainside crash. Tail section is hung over embankment to give huge cargo plane every inch of runway for take-off.

Upper 1957 photo is Morrison-Knudsen's work horse. The four-engine LB30, the civilian forerunner of the famed Liberator bomber of World War II. The present LB30 replaces a similar airplane that was retired from service in late 1951 after flying more than 5,000 hours on the Seattle-Alaska run and throughout Alaska. The big ship and its successor have landed on practically every airstrip in Alaska to deliver men and materials to M-K jobs. Powered by four 1200-hp engines, the "LB" cruises at 200 miles per hour and will carry 34 passengers or 14,000 pounds of freight.

Morrison-Knudsen Archives

Northern Consolidated

Three photos graphically portray unusual and varied payloads of newer, larger and faster aircraft serving the 586,400 square miles of the State of Alaska.

Opposite: In the eve of the 1960s, dog team travel of the past meets new turbo-prop aircraft of the present. The introduction of jet power becomes an aviation milestone in Alaska.

Northern Consolidated

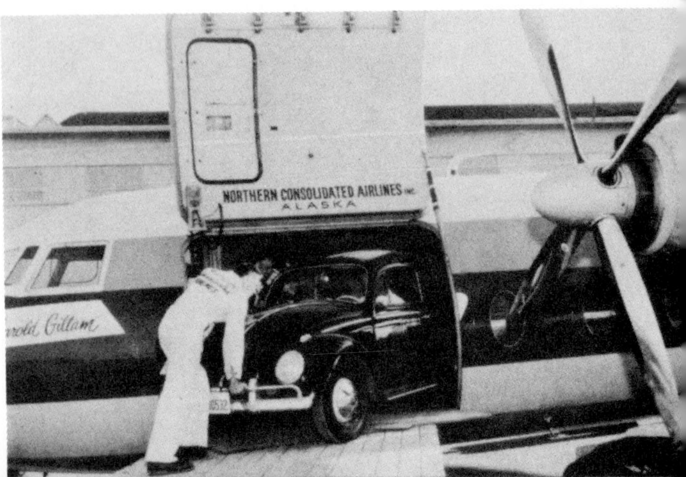

Northern Consolidated

Hayes Collection

Maj. Gen. Frank A. Armstrong, Jr., Commander of Alaskan Air Command, right, greets Secretary of State John Foster Dulles and John M. Allison at Anchorage in 1956. Armstrong and Col. W. Preston Corderman, left, toured Dulles through AAG headquarters and Elmendorf Air Force Base facilities. Fast, long-distance air travel made such inspections frequent.

WORLD AIRLINES JUNCTION

Alaska Is Gateway of Air Age

CHAPTER 11

DURING HIS COURT-MARTIAL HEARINGS held in 1935, pioneer military aviation enthusiast Brigadier General William "Billy" Mitchell said, "Alaska is the most central place in the world for aircraft, and this is true either of Europe, Asia or North America. I believe in the future. He who holds Alaska will hold the world, and I think it is the most strategic place in the world."

Long distance flight experience gained during World War II with heavy bombers and in the ferry command made it apparent that with the cessation of hostilities, world air travel would enter a new dimension. Using soon-to-be-built faster, long range transports, airlines would fly great circle routes; the shortest global distance between major world cities. In the northern hemisphere, the shortest distance between Asia, Europe and the United States is over the North Pole.

Soon industrial companies from all over the United States and elsewhere joined the growing Alaska air traffic with their executive and utility aircraft.

Anchorage, with the largest number of based aircraft of all descriptions, soon became the aviation center of the Territory and by serving air carriers from many nations, as well as the United States, was recognized as the "Air Crossroads of the World."

By the late 1950s, Anchorage International Airport, Elmendorf Air Force Base, Lake Hood Seaplane Base and Merrill Field had created a maze of high density air traffic patterns which became one of the world's busiest aviation centers.

First jet flight to Anchorage was Pan American World Airways' Boeing 707-VIP flight in 1960. On its New York to Paris route, Pan Am was the first airline to put the giant Boeing into service on October 26, 1958.

Boeing Co.

Hayes Collection

Post-war surplus military aircraft placed on the open market offered thousands of ex-military and non-military pilots an opportunity to own their own airplane or airline. A huge crop of non-certificated air carriers sprung up. They flew what amounted to charter flights, and at times over routes of scheduled carriers; often times less than half their fare. These operators, known as non-scheduled airlines (or non-skeds), created stiff competition viewed unfair by owners and stockholders of certificated carriers, who were required to have millions invested in facilities, equipment and personnel to meet established schedule. An air travel war ensued, which had its effect on air travel to and from Alaska.

NON

All Photos Gordon Williams Collection

Author's Note: Possible confusion of airline names in the following chapter may be clarified with the following: Alaskan Airways was Fairbanks-based from 1929 until merger to Pacific Alaska Airways in 1932. Alaska Airlines was transition name of Anchorage-based Alaska Star Airlines after World War II. Wien Alaska Airlines was transition name of Wien Airways. None of the above were inter-connected.

SKEDS

Mills Collection

Alaska Airlines

Upper: Founders of Star Air Service: (L-R) Jack Waterworth, Charley Ruttan and Steve Mills in 1932. Right: Pilot Estol Call of L. Mac McGee's McGee Airways in 1933. A Star-McGee merger became parent company for Alaska Star Airlines in 1937.

Alaska Airlines

Above: 1943 photo of Star Air Lines Ford Trimotor which served all of Alaska for years. First Ford acquired by the airline in 1935 fell through the ice at Lake Spenard while being fitted for skis during the winter of 1935-36.

ALASKA

Alaska Airlines

Right: Twin-engined Douglas DC-3 Starliners joined the airline beginning in 1943. They were flown on trunk routes between Anchorage and major points throughout Alaska. The name was changed from Star Airlines to Alaska Star Airlines that same year.

Above are veteran AA pilots (L-R) Capt. Wilfred (Sonny) Lund, 1st Officer Arthur L. Clune and Capt. Don Culver. Upper right: Capt. Clark Cole greets passengers boarding his DC-4 flight. Service to Seattle began in 1946 and temporarily certificated in 1951.

AIRLINES

Alaska Airlines was all but defunct in 1957 when World War II Navy veteran Charles F. Willis, Jr. bought the line and with DC-6, above, built the airline up to the successful "Golden Nugget" service from Alaska points to the South with Boeing 727.

All Photos Alaska Airlines

Gordon Williams Collection

One of many bush operated twin-engine Cessna Bobcats of the fast growing NCA. Experience and shrewd operations in a highly competitive business skyrocketed NCA from a small airways in 1946 to Alaska's most up-and-coming airline in the early 1950s.

NORTHERN

Upper: Raymond Petersen founded Petersen Flying Service in 1935. With merger of Jim Dodson Flying Service, Bert Ruoff's Bristol Bay Flying Service and Johnny Wakatka Flying Service, Northern Consolidated Airlines was formed with Petersen as president.

Northern Consolidated

Remote interior station of the fast-growing Northern Consolidated Airlines.

NCA operated four DC-3's from June 1949 to October 1959 throughout all of Alaska, in conjunction with their smaller bush operations. Left is retirement photo of first DC-3, "Old 748." The honorable "gooney birds" were replaced by the new turbo-prop Fairchild F-27.

Below: Indian boy passenger wrestles with his gear following a NCA flight into interior Alaska. Fairchild Hiller FH 277 shown here, is one of several modern, smaller aircraft NCA put into bush operations in 1962.

CONSOLIDATED AIRLINES

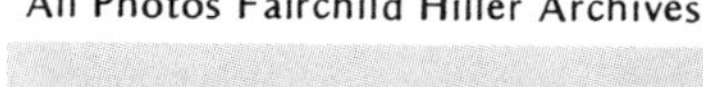
All Photos Fairchild Hiller Archives

Below: Before NCA merger with Wien Airlines in 1968, with Petersen as president, F-27 was basic aircraft which served well in the wilderness of Alaska.

Northwest Airlines' president, Croil Hunter, is met by Anchorage airport manager Clay Shupe, right, as Hunter arrives from States on trip to the Orient in November 1947.

NWA Douglas DC-4, used on airlift to Alaska during World War II when not assigned to regular company routes. Twenty-seven NWA pilots flew in Alaska, either in the military of ATC, or both.

All Photos Northwest Airlines Archives

NORTHWEST

Left: Crew members on first NWA Orient flight from Twin Cities Wold Chamberlain Field. (L-R) Jerry Koerner, radio; Donald Rector, engineer; George Bickel, navigator; Virgil R. Carlson, purser; Larry Horner, 1st officer; Evelyn Currie, stewardess; Ed LaParle, captain. Ceremonies below.

Above: Fleet of NWA Douglas airlines on Anchorage stopover, enroute to and from the Orient in 1949.

Hayes Collection

ORIENT AIRLINES

Passengers disembark at "Air Crossroads of the World," Anchorage, from one of NWA Stratocruisers that served the growing airline throughout the 1950s. Below: Present-day equipment, Boeing 747 jumbo jet.

Boeing Co.

Lewis Photo

Above: Pioneer bush pilot Joe Crosson first flew in Alaska for the Fairbanks Airplane Company in 1925.

Photo by Noel Wien

Above is Fairchild 71 of Alaskan Airways of Fairbanks, founded by Joe Crosson and Carl Ben Eielson in 1929.

Below: Fairchild 71 of Pacific International Airways of Anchorage, which merged with Alaskan Airways in 1932 to form Pacific Alaska Airways, with headquarters at Fairbanks.

Mills Collection

PAN

Pacific Alaska purchased this all-metal twin-engine Lockheed Electra in 1935, and operated out of Fairbanks.

Gordon Williams Collection

PAA, a subsidiary of Pan American Airways, inaugurated Fairbanks to Juneau flights in 1935. Juneau-Seattle service was first flown with four-engine flying boats in 1940. Later, Lockheed Lodestars took over.

Gordon Williams Photo

Above: PAA DC-3 takes off from Boeing Field during World War II. Under contract to the U. S. Navy, PAA flew out of Seattle on Alaska ports of call as far north as the Aleutians.

Monsen Collection

In 1946 photo above, PAA Capt. Al Monsen, center, with his son, 1st Officer Wesley Monsen, left, pilot-engineer Bob Bedinger, right, and unidentified stewardess, before take-off from Seattle to Juneau flight.

AMERICAN

In 1958, PAA was first airline to accept delivery of famed Boeing 707, similar to one pictured below. In addition, the line currently flies 22 Boeing 747's on their world-wide routes.

PAA Photo

Robert C. Reeve established Reeve Airways in 1932 at Valdez, Alaska, and presently holds the position of president of that airline; now known as Reeve Aleutian Airways.

With Anchorage and Fairbanks as bases of operations, Reeve used Fairchild monoplanes and Boeing 80-As, pictured above, to fly men and supplies for the Army in World War II.

Gordon Williams Photo

All other Photos Courtesy of Reeve Aleutian Airways

REEVE

Above is Gene Craig, Reeve's private secretary for 15 years and manager of Passenger Reservations and Service. Reeve attributes success of the line to his loyal employees.

Capt. William R. Borland, manager of Air Operations, has 25 years and 20,000 flying hours in Alaska and the Aleutians. In WW II, he was a Marine pilot in the South Pacific.

Reeve originally had four DC-3 airliners serving the Aleutians. Early 1950 photo below of Merrill Field based N-19906 is still operational with over 30,000 flying hours.

Above: Richard Reeve, oldest son of Reeve, is Administrative Vice President and was formerly a line captain.

ALEUTIAN AIRWAYS

Above: Air Force motion picture crew and Reeve personnel with backdrop of DC-4 and Anchorage International control tower. Film was Air Force documentary.

Right: Dave and Janice Reeve, pilot and stewardess of Reeve Aleutian Airways. Below: Lockheed L188C Electra serves nine Aleutian ports of call.

Above: Roberta Reeve, former RAA stewardess; now married to bush pilot Don Sheldon of Talkeetna.

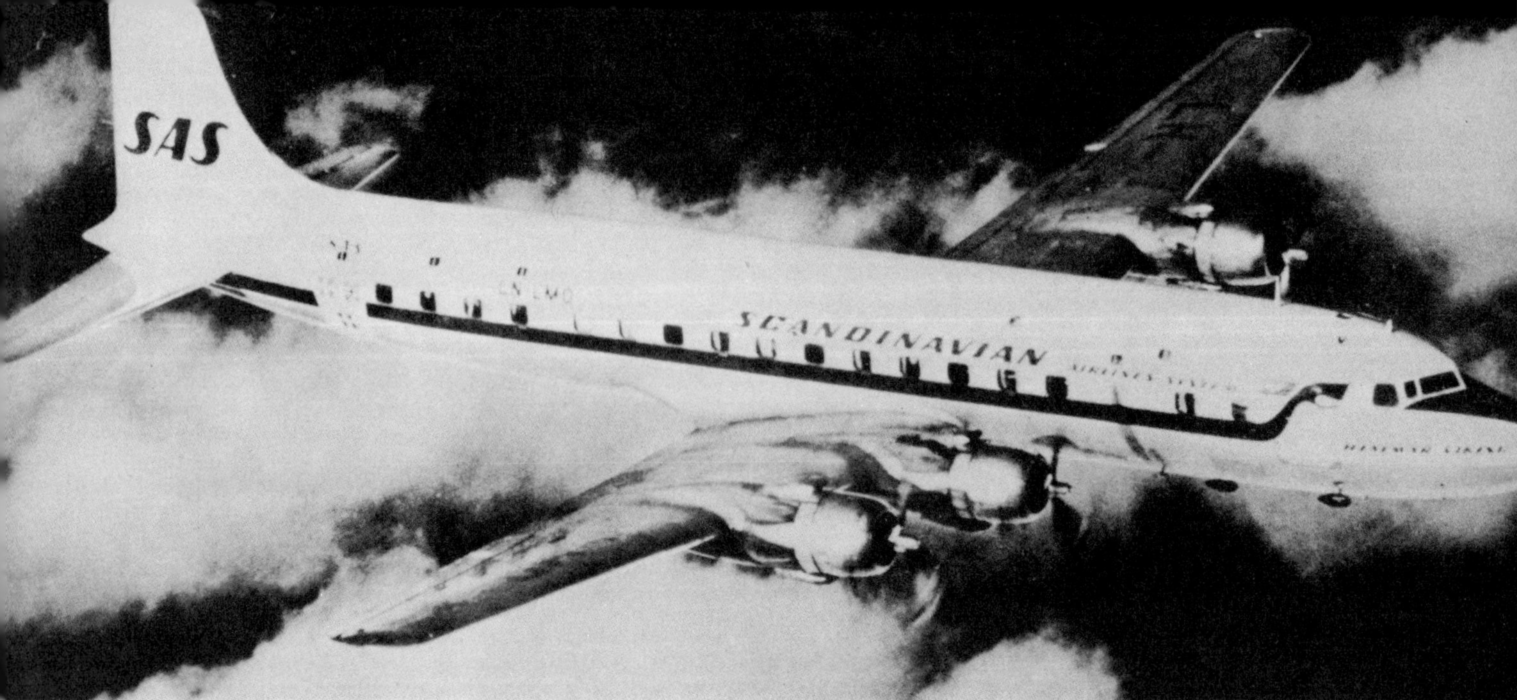

SAS Photo

Scandinavian Airlines System inaugurated the world's first transpolar service in 1954 between Copenhagen and Los Angeles. In February 1957, a twice-weekly flight from Copenhagen to Tokyo with a stop at Anchorage. The Douglas DC-7C shown above.

SCANDINAVIAN

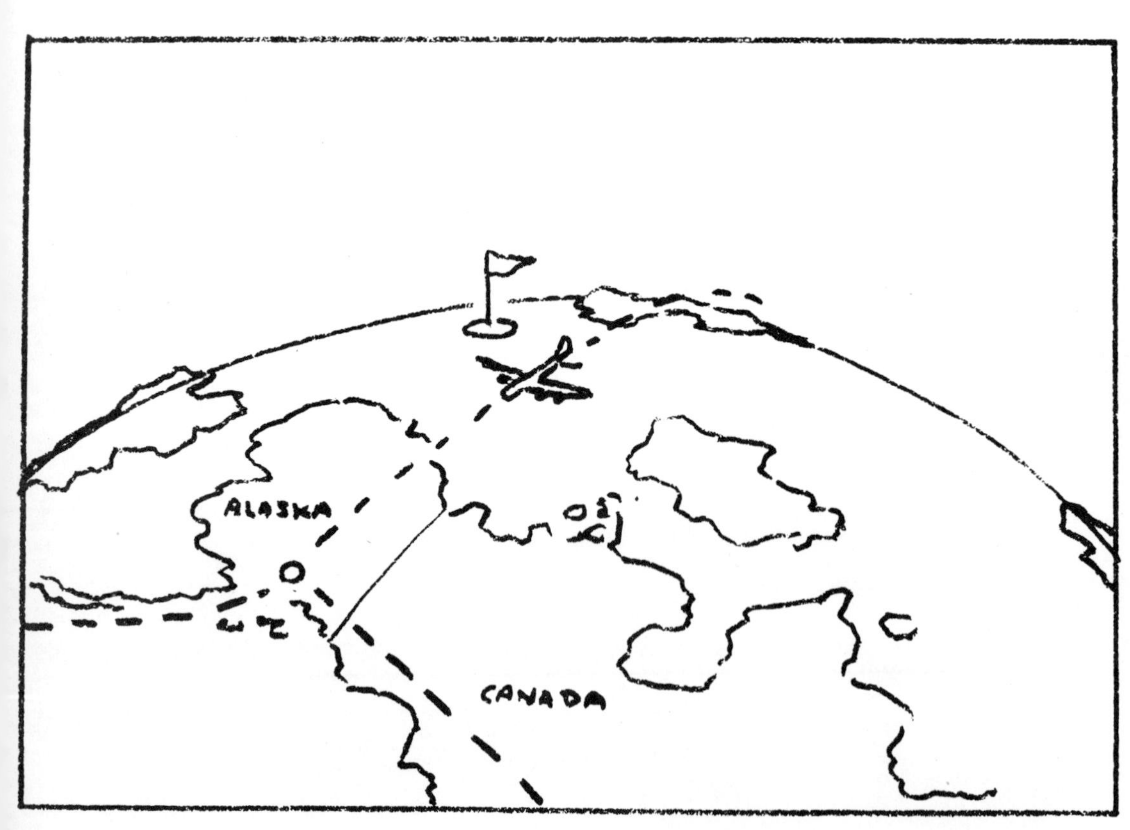

SAS Photo

SAS twin-jet French Caravelle maintains scheduled flights between Denmark, Sweden and Norway. Shown above is the Finn Viking.

AIRLINES SYSTEM

With a stop at Anchorage International Airport, SAS offers over the North Pole flights to the Orient and United States using stretch-out Douglas DC-8 Super Fan aircraft. Below is the Leif Viking.

SAS Photo

Left upper: Arthur G. Woodley, who founded Woodley Airways in 1930. Above: One of several Travelair 6000s used throughout the 1930s and 1940s. Left: Company name was changed to Pacific Northern Airlines in 1945; later DC-3s were acquired.

WESTERN

Above: First DC-4 at Elmendorf AFB. In 1951, PNA routes extended to Seattle.

Below: Longtime familiar "Connie" which PNA operated Seattle-Anchorage and points north in the 1950s and 1960s.

Western Airlines Photos This Page

Right: Fast, trim, little Lockheed 12 which replaced twin-engine Cessnas serving key points out of Anchorage. Right lower: Dave Kellogg, now a captain on Western jets, joined that airline in 1947.

Kellogg Collection

AIRLINES

Below: Pacific Northern merged with Western Airlines in 1967. Fleet of Boeing 720s like one shown below serves the Seattle-Anchorage route.

Western Airlines Photo

Photo by Noel Wien

Above: Noel Wien and Hisso Standard used to start operations in Nome in 1927. Wien first came to Alaska in 1924.

Noel Wien Photo

Above: Jim Dodson flying Wien Bellanca over Noel Wien's Fairchild 71, being loaded.

Noel Wien Collection

Above: Three Wien brothers. (L-R) Noel, Sigurd and Fritz in 1946, soon after acquiring two DC-3s. A fourth Wien brother, Ralph, was killed in 1930.

WIEN

Below: Impressive line-up of Wien C-46s and DC-3s during the 1950s.

Wien Consolidated

Above: Commuting high school and college students prepare to load aboard Wien twin Otter.

ALASKA AIRLINES

Above: Typical scene in Alaska as Wien bush plane meets fast F-27 turbo-prop airliner. Wien became one of that state's largest air carriers in the 1960s.

Photos This Page Wien Consolidated

Wien Airlines and Northern Consolidated merged in 1968. Below is one of Wien Consolidated's Boeing 737 transports used throughout Alaska.

Boeing Co.

Contrasting the new with the old, a quarter-century of air progress is shown here in this 1958 photo. The new, the Boeing 707 jet; the old, the eight-passenger "Pilgrim" built by American Airplane & Engine Company (now Fairchild Hiller) for American Airways (now American Airlines) in 1931. Replaced by American in 1937 and operated by Alaska Airlines since 1939, the 100-mph antique was sold to the Salmon River Flying Service in 1957. Thence followed duties as a firefighter for the State of Washington and in 1964 was sold to Marine Packing Co. of Seattle for salmon hauling in Alaska. The nostalgic bird was reacquired by Alaska Airlines in 1970; in 1971, on its 40th birthday, it was sold again.

U. S. Air Force

Flyers and aircraft from Jenny to jet have headlined over sixty years of aviation progress, none of which could be achieved without the visionary designer, analytical engineer, dedicated mechanics, far-sighted communications technician, super-human air traffic controller and thousands of support personnel who represent the backbone of flight.

NWA Photo

Right: Pretty stewardess Charmalee Prentice flew Northwest Orient Airlines' Tokyo flights following two years flying with United Airlines. Miss Prentice symbolizes a vital air crew profession that has served passenger carrying flights for over 40 years. The first stewardess was Ellen Church.

Alaska Senior Senator Bob Bartlett, right, receives congratulations on Alaska's Statehood from Northwest Airlines' vice-president Don King. Senator Bartlett, a protege of, and in 1944, successor to Alaska Delegate to the U. S. Congress Anthony J. Dimond, was actively engaged in pioneer development of air commerce in Alaska and, later, championed Alaska's Statehood.

Northwest Orient Airlines

U. S. Air Force

AIR DEFENSE SCREENING – The showcase for signs of air attack against Alaska and rest of North America is this display board in the combat center of the Alaska Region of the North American Air Defense Command (NORAD). Radar eyes throughout the State provide Alaskan NORAD Region commander Lt. Gen. Robert A. Breitweiser and deputy commander Maj. Gen. Thomas E. Moore with up-to-the-minute picture of air activity in the region.

U. S. Air Force

Bibliography – published works

Arnold, Maj. Gen. Henry H. *Our Air Frontier in Alaska.* National Geographic LXXVIII (Oct. 1940), 487-540.

Chambers, John R. *Arctic Bush Mission.* Seattle: Superior, 1970.

Clarke, Basil. *Polar Flight.* London: Ian Allan, 1964.

Day, Beth. *Glacier Pilot.* The story of Bob Reeve. New York: Holt, Rinehart and Winston, 1957.

Ellis, Robert E. *What-No Landing Field.* Haines, Alaska: Chilkat Press, 1969.

Garber, Paul E. *National Aeronautical Collections.* Washington, D.C.: Smithsonian Institution, 1965.

Garfield, Brian. *The Thousand-Mile War:* World War II in Alaska and the Aleutians, New York: Doubleday, 1969.

Keithahn, Edward L. *Alaska for the Curious.* Seattle: Superior, 1966.

LeMay, Gen. Curtis E. with McKinlay Kantor. *Mission with LeMay.* New York: Doubleday, 1965.

Okumiya, Masatake and Jiro Horikoshi with Martin Caidin. *Zero.* The Story of Japan's Air War in the Pacific: 1941-45. New York: Ballantine Books, 1971.

Potter, Jean. *Alaska Under Arms.* New York: MacMillan, 1942. *The Flying North.* New York: MacMillan, 1947. *Flying Frontiersmen.* New York: MacMillan, 1956.

Ross, Capt. James L. USAF. *History Shemya Army Air Force Base.* Anchorage: Alaskan Air Command, 1969.

Spring, Bob and Ira with Byron Fish. *Alaska.* Seattle: Superior, 1965.

United States Army. *The Army's Role in the Building of Alaska.* Public Information Office, Headquarters U. S. Army, Alaska, 1969.

Wachel, Pat. *Oscar Winchell: Alaska's Flying Cowboy.* Minnesota. Dennison and Co., 1967.

Wigton, Don. *From Jenny to Jets.* New York: Bonanza, 1963.

CONTRIBUTORS TO POST-WAR SEGMENT

Alaska Wing Civil Air Patrol:
- Lt. Col. Charles W. Burnette, Special Projects
- CWO Shirley Fletcher, Administration
- Capt. Paulette Poyneer, Information Officer
- Major Allen H. Shewe, Director of Information

Gene Craig, Private Secretary, Reeve Aleutian Airways

Dexter Cushing, Public Relations, Boeing Co.

G. Delisle, Chief, Photos. Public Archives, Canada

Irmgard Dennis, Aerospace Writer, McChord AFB

James H. Dilonardo, President, OX-5 Washington Wing

Capt. Vincent Doyle, Historian, Northwest Airlines

Leif Eie, Mgr., Seattle, Scandinavian Airlines

General Services Administration, National Archives and Records Service

Lt. Col. Gerald M. Holland, USAF, I.O., Washington, D.C.

LCDR J. J. Jamrogh, USN, Photo Lab, Washinton, D.C.

Capt. Maurice Keating, Jr., Western Airlines

C. O. Kempton, Chief, Airports Div., FAA, Anchorage

Harry O'Hara, Property Off., Alaska Airlines

Robert Parke, Publisher, Flying Magazine

Theron Rinehart, Pub. Rel., Fairchild Hiller

Peter Robertson, Public Archives of Canada

Robert F. Smith, Pub. Rel., Morrison-Knudsen Co.

Torval "Toby" Steen, U. S. Dept. of Fisheries

Peter Tornquist, Mgr., L.A., Scandinavian Airlines

Charles F. Willis, Jr., Board Chairman, Alaska Airlines

Gordon Williams, Public Relations, Boeing Co.

INDEX